OTHER TITLES IN THE SERIES

Alternative Kilns
Ian Gregory

Ceramics and Print
Paul Scott

Coiling
Michael Hardy

Ccramics with Mixed Media
Joy Bosworth

Colouring Clay
Jo Connell

Crystalline Glazes
Diane Creber

Dry Glazes
Jeremy Jernegan

The Electric Kiln
Harry Fraser

Glazes Cone 6
Michael Bailey

Handbuilding
Michael Hardy

Impressed and Incised Ceramics
Coll Minogue

Kiln Building
Ian Gregory

Large-scale Ceramics
Jim Robison

Lettering on Ceramics
Mary White

Low Firing and Burnishing
Sumi von Dassow

Oriental Glazes
Diane Creber

Paper Clay
Rosette Gault

Porcelain
Jack Doherty

Raku
John Mathieson

Resist and Masking Techniques
Peter Beard

Setting Up a Pottery Workshop
Alistair Young

Single Firing
Fran Tristram

Slipcasting
Sasha Wardell

Stoneware
Richard Dewar

Soda Glazing
Ruthanne Tudball

Throwing Pots
Phil Rogers

Wall Pieces
Dominique Bivar Segurado

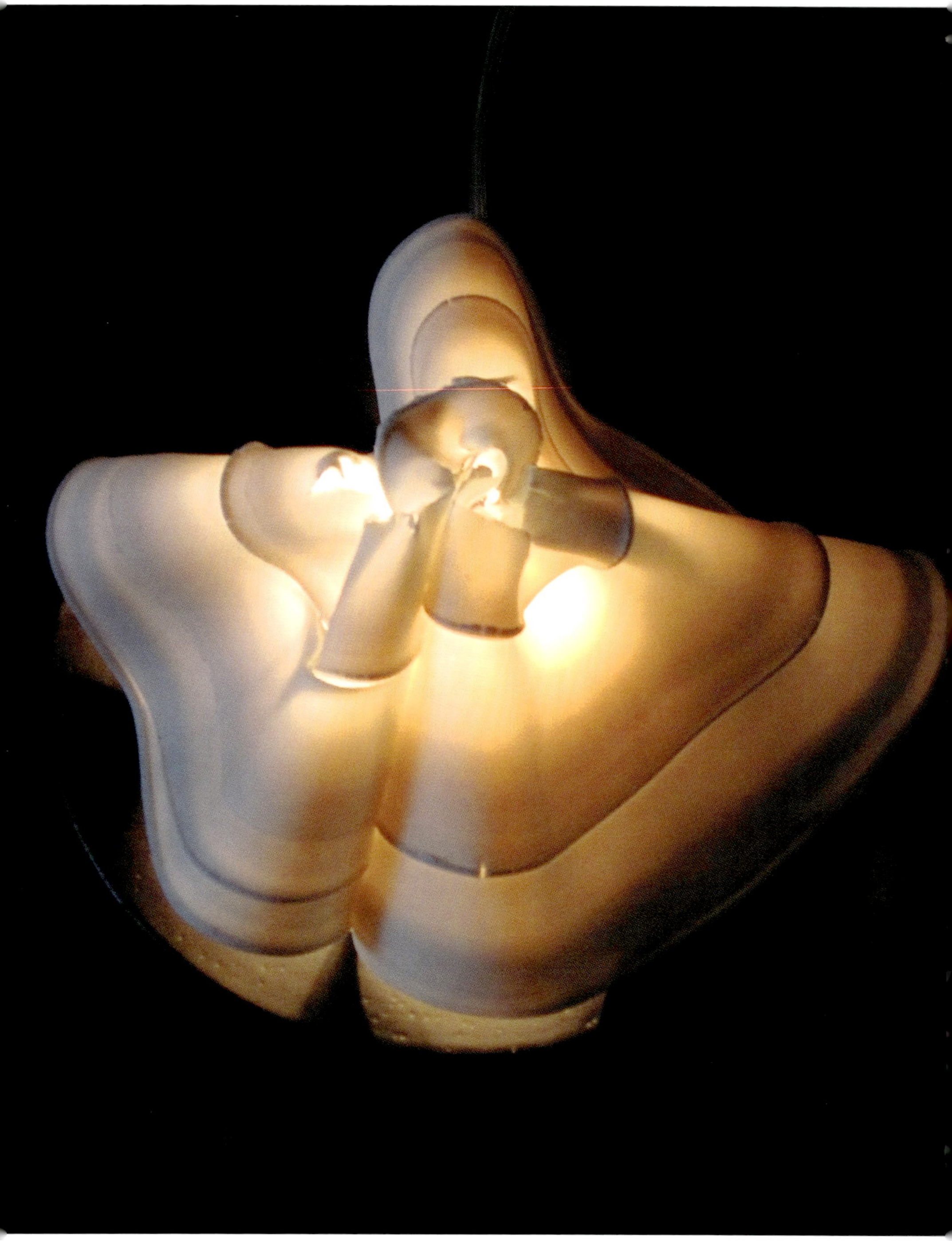

CLAY, LIGHT & WATER

Margaret O'Rorke

A & C Black • London

University of Pennsylvania Press • Philadelphia

Margaret O'Rorke.

FRONTISPIECE *Sea Frills*, Margaret O'Rorke, 2002. D: 22 cm (8¼ in.). Thrown porcelain. *Photo: Margaret O'Rorke.*

COVER (FRONT) *Two Flying Forms*, Margaret O'Rorke, 2008. L: 40 cm (15¾ in.). Thrown, re-formed and joined porcelain. *Photo: James Crabbe.*
COVER (BACK) *Bubbles*, Margaret O'Rorke, 1998. D: 50 cm (20 in.). Thrown porcelain. *Photo: John Woroniecki.*

First published in Great Britain in 2010
A & C Black Publishers Limited
36 Soho Square
London W1D 3QY
www.acblack.com

ISBN 978-0-7136-8487-2

Published simultaneously in the USA by
University of Pennsylvania Press
3905 Spruce Street
Philadelphia, Pennsylvania 19104-4112

ISBN 978-0-8122-2099-5

CIP Catalogue records for this book are available from the British Library and the US Library of Congress.

Typeset in 10 on 12½pt Photina
Book design by Susan McIntyre
Cover design by Sutchinda Thompson

Printed and bound in China.

This book is produced using paper that is made from wood grown in managed, sustainable forests. It is natural, renewable and recyclable. The logging and manufacturing processes conform to the environmental regulations of the country of origin.

Contents

Foreword

When I first saw the work of Margaret O'Rorke in an exhibition about light, I was dazzled. Her high-fired translucent porcelain forms were part sculpture, part sci-fi fantasy.

I became a fan. When I went to interview Margaret at her 15th-century 'hall house' outside Oxford, it was exactly what I'd hoped for. The timber-framed house, with its stone floors, exposed beams and open fireplaces, is the perfect backdrop for her contemporary porcelain lights. Vessels resembling seedpods and cocoons cascade down the plaster walls. There are wavy wall sculptures that cast unusual shadows, and several orb-shaped works, glowing with light, appear to float across the interior.

O'Rorke is a force of nature. Now in her late sixties, she seems impossibly youthful, with her hair cut in an urchin crop, she often wears a pair of extremely fashionable sports shoes designed by her son. 'I love being the age I am now,' she laughs.

For the past 26 years O'Rorke has been creating and exhibiting exquisite thrown forms that give light. 'I love the translucency of fine, high-fired porcelain,' she says. 'Light bounces off most pottery forms, but I wanted my pots to emit light.'

What makes O'Rorke's work so intriguing is its organic, ethereal quality, epitomized in her use of winglike pleats. She often finds herself dreaming up new pieces to illuminate the individual spaces of the house. 'The height of the open roof space on the first floor has enabled me to experiment and build large-scale installations; porcelain chandeliers can be suspended from the beams or hung above the balcony.'

O'Rorke trained as a painter and ceramicist but had to give up when she married at 21. 'I didn't start potting until I was 40 and by then the whole ceramics world had changed. I was still making quite traditional thrown pots, but feeling very ill at ease. I wanted something more, but I had no idea what.' She began experimenting: 'I realised I could have such adventures using the light and the clay.' She admits it was 'agony' learning to put electrics into her pieces. 'I never ever want to see any light fittings. I hate seeing the cord – it has to look magical.' Today her installations, including chandeliers, wall pieces and fountains, all involve complex engineering, both electrical and mechanical, born of that irrepressible urge to see the clay lit up.

O'Rorke is also a collector. She loves to display fellow artists' work throughout the house's interior. Her treasures include tableware by her great friend, the late Joanna Constantinidis; a jug and bowl by Lucie Rie; and sake dishes and jugs by the Japanese master potter Koie Ryoji. So it comes as no surprise that writing this book, she is keen to celebrate other contemporary artists using clay. As she explores what draws a ceramic artist to include artificial light within their work, she brings us up to date with the work of studio potters such as Sweden's Mia Göransson, who creates sculptural bone china bud and leaf forms, and Japan's Kurokawa Kenkichi, who makes cast wall lamps. These, she shows us, are artists who also 'talk with their hands'.

Liz Hoggard
Arts writer and columnist, *Evening Standard*

Chapter 1

Clay and light: a brief introduction

Ceramics may have been discovered thousands of years ago when meat was cooked on an open wood fire and the fat caught alight as it melted into the flames, causing the clay beneath the red-hot fire to become hardened by the heat. Certainly, the ingenuity, imagination and craftsmanship of the potter played a huge part in the early development of light created for use during the hours of darkness.

Aristotle (384–322BC) wrote in his *Meteorologica* that the whole terrestrial region was composed of four 'bodies': fire, air, water and earth. Humanity's early struggles and ultimate ingenuity with fire, with light and with the earth's basic elements are akin to the potter's craft of clay, water and fire. Potters have embraced these elements for thousands of years, from the making of early fire and clay lamps to the advanced technology available today. We know the sense of calm and intimacy that candlelight or fire creates for a gathering of friends, how it draws people together, and this has not changed over time.

A glance at the history of clay lamps over the last 3,000 years shows that forms developed very slowly. Human ingenuity has gradually revealed itself over thousands of years, in the discovery and development of the complex scientific processes needed to create artificial lighting.

A simple open dish with a floating wick is the earliest recorded type of clay light, similar to the low-fired oil lamp (see photo below) and still to be found in remote parts of the world where electricity is not available (see photo p.9). This tried and tested solution was developed in the age of the Mycenaean civilization, around 1500BC.

In Egypt, China and Japan, and more recently in Africa and India, a saucer was placed on a stand or joined to a stand to give a greater spread of light. The Greeks and Romans from the 6th to the 4th centuries BC developed this idea, curving the rim to prevent oil spillage, providing a support for the wick and adding a handle, making the lamp more easily transportable.

Low-fired oil lamp with floating wick.
Photo: James Crabbe.

Women preparing oil lamps in India, 2008. *Photo: Lucy O'Rorke.*

The Greek lamps were mostly thrown on a potter's wheel, while the Romans developed moulds from gypsum (calcium sulphate). Spout lamps (see photo on opposite page) were developed and used in the Mediterranean well into medieval times. A rich array of clay lamps was used for religious ceremonies. In time, large, multi-wicked, freestanding and hanging oil lamps became a feature of huge banquets, suspended around the hall.

The type of fuel used for these early lamps depended on the location, environment and natural resources available. In the Arctic, eskimos used whale blubber and fish oil to create light, as well as to cook with, in their igloos. Wicks were made from dried moss gathered in spring as the ice melted. Moving south to more temperate climates, animal fat, olive or seed oils were burnt. Cotton and other porous materials and plants such as reeds and papyrus were used for wicks. 'The heat created once the wick is lit allows capillary action to draw the fuel to the flame, which ignites the gas creating the light' (see Faraday's *The Chemical History of a Candle*, J.M. Dent & Son.)

Clay lights using wicks began to diminish from the mid-15th century onwards, and tallow (rendered and refined animal fat) and beeswax became the basis of candle-making. This was followed centuries later by paraffin, and the paraffin or kerosene lamp radically changed the design and function of the flame for creating light. There was also a finer level of engineering and design of the wicks, which increased both the efficiency and the safety of the lamps. Kerosene is still used extensively for lighting around the world where gas and electricity are not available.

The court of a thousand clay lamps, Nagur Fort, Rajasthan. *Photo: James Crabbe.*

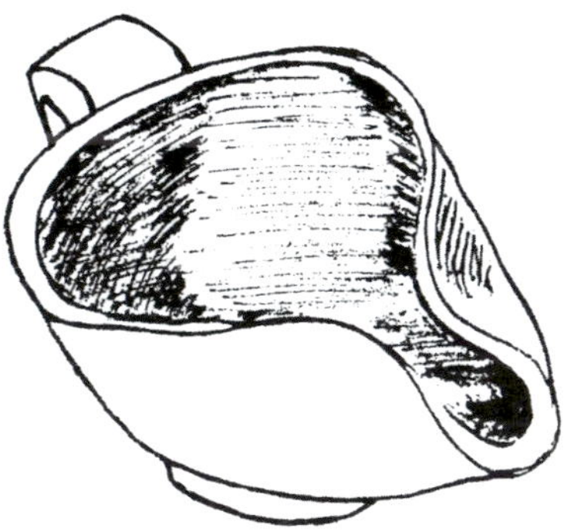

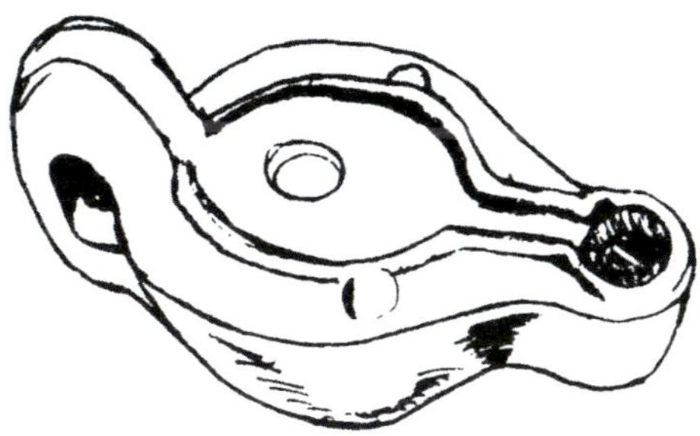

Other clay lamps
ABOVE The shell lamp used by the Ancient Egyptians, Chinese and Japanese.

TOP RIGHT Greek and Roman lamp, 6th century BC.

RIGHT The spout lamp derived from the Roman lamp. *Drawings by Kozo Hida, courtesy of the Pitt Rivers Museum, Oxford.*

23 individual oil lamps used for ceremonial occasions, Kabylie, Algeria. Pitt Rivers Museum, Oxford. *Photo courtesy of the Pitt Rivers Museum, Oxford.*

On 28th January, 1807, gas lighting was introduced in Pall Mall in London. Pall Mall was the centre for gentlemen's clubs in the city, and the lighting afforded an important measure of safety. In 1812, the London and Westminster Gas Light and Coke Company became the first gas company in the world. Gas lighting developed rapidly, but by the late 19th century it was superceded by electricity as the main fuel for lighting. This was the beginning of the lighting industry as we know it now, and it changed the part ceramics was to play in the future of lighting. At that time ceramic vessels for lamp bases in the home would hold a conventional down light shade of silk or paper.

In the images opposite you can see a very simple oil lamp I made myself, which was inspired by the oil lamps at the Pitt Rivers Museum in Oxford. I researched different types of oils and wicks and found that thick cotton yarn plaited loosely made a good wick using sunflower or olive oil. Careful trimming of the wick can produce a clean flame. There is a familiarity here for the potter accustomed to tuning their kiln to give a clean reduction atmosphere.

Contemporary creative artists using clay can consider how the nature of the fired material combined with a powered light source can add to the content of the space it will inhabit. What we want the light to do, where it will be placed and

RIGHT A contemporary oil lamp by Margaret O'Rorke, using sunflower or olive oil and plaited cotton wicks inspired by lamps in the Pitt Rivers Museum. Ht: 34 cm (13 in.) (approx.). *Photos: James Crabbe.*

how it will work are some of the questions the lamp-maker needs to consider.

As a studio potter making thrown domestic ware, I stumbled across the idea of combining the translucency of porcelain with electric light (see photos pp.12–13). At the time, I thought I was the first potter to think of the idea. To begin with I struggled to master throwing porcelain vessels tall enough to house the light fittings and thin enough to be translucent, always wanting the feeling of energy and the instant touch of my hands to be revealed and transmitted by the light.

Since then, I have discovered that others were using light with porcelain earlier than me. Rudolf Staffel, an American ceramic artist and teacher born in San Antonio Texas in 1911, is well documented for his *Light Gatherers* (see p.14) whose light shines through the body of the clay and into the surrounding space.

A few of his pieces, I believe, did contain an electric light bulb, although the majority were displayed with a light above shining into the centre of the vessel. This highlighted the translucent qualities of both the porcelain and the making process, and greatly simplified the technical difficulties by removing the problem of having to conceal the light fitting, lamp and flex.

An early thrown vessel form by Margaret O'Rorke, 1987, showing oxidation (left), and reduction (right). Ht: 32 cm (12½ in.). *Photo: Margaret O'Rorke.*

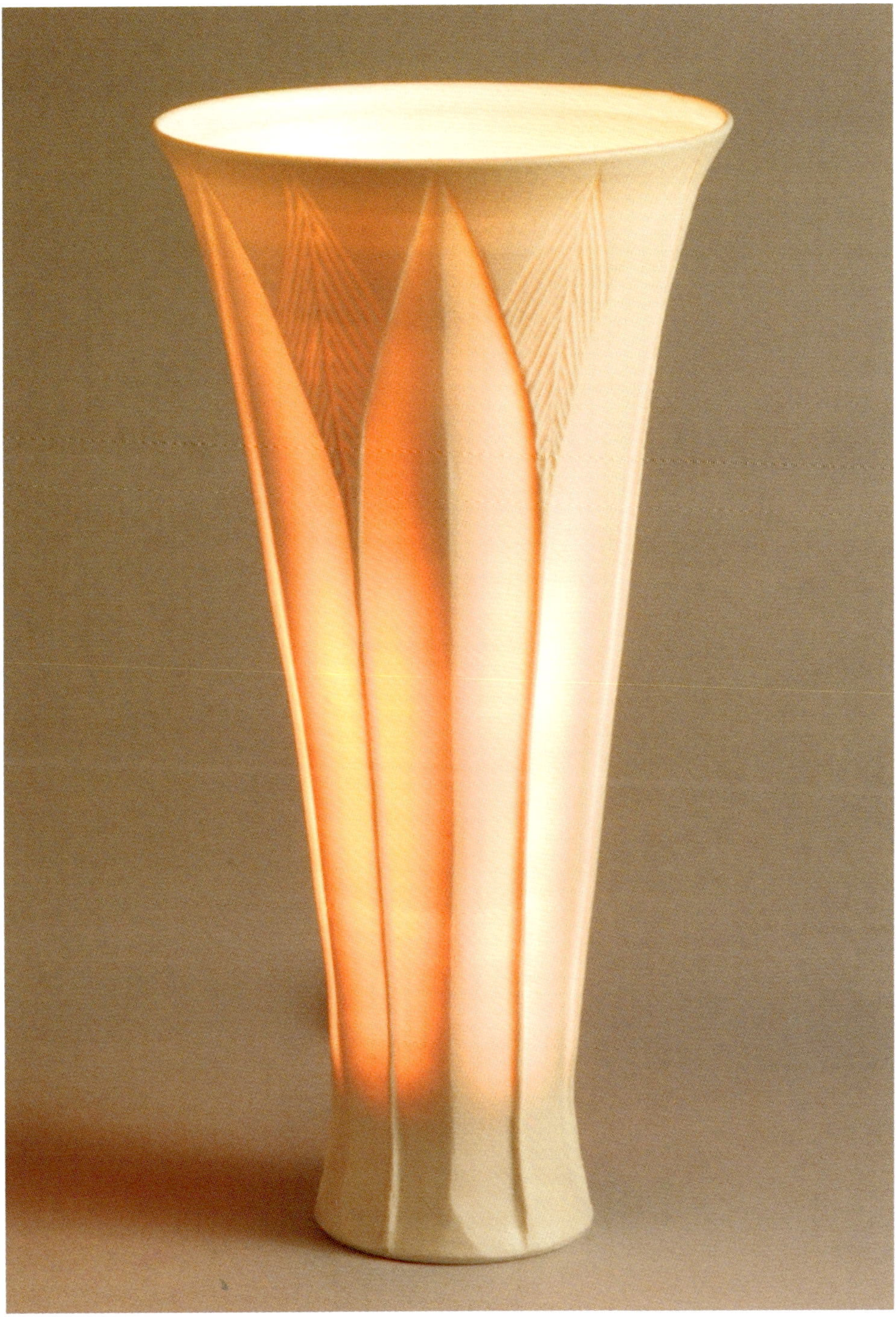

Thrown and carved trumpet by Margaret O'Rorke. Ht: 35 cm (13½ in.). *Photo: Sherwin Pearson.*

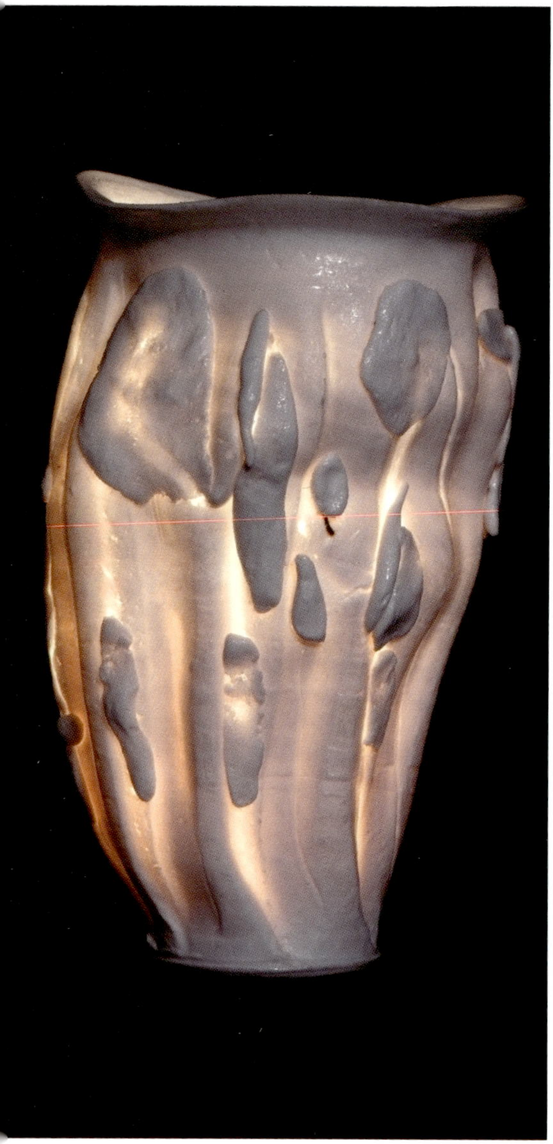

Light Gatherer, Rudolf Staffel, 1981, USA. Wheel-thrown porcelain vase with porcelain additions. Ht: 20.3 cm (8 in.), D: 12.8 cm (5 in.). *Photo: courtesy of V&A Images, with permission from the estate of Rudolf Staffel.*

Poul Johannes is a Danish ceramic artist who has explored a wide range of ceramic materials and processes. He became Design Director for Porsgrund Porcelain Factory and went on to become a professor at the Oslo National Academy of the Arts. More recently, he held the position of Director of the International Ceramic Research Centre in Denmark. In the late 1970s and early '80s he created some Danish-style simple cast-porcelain shades. He used porcelain slip from Porsgrund, reduction fired to 1330°C (2426°F). These pieces were never put into production, but like myself he has produced them from time to time in his studio (see images opposite on p.15).

Creative artists using translucent clay can now embrace the current lighting technology and consider how the nature of the translucent fired clay combined with a light source can transform a contemporary interior, creating a calm and peaceful atmosphere. Skilled contemporary lighting designers can use the scientific developments that have evolved over the centuries to exploit all aspects of lighting to their advantage. For the ceramic artist the opportunities have never been greater.

Porcelain lamp, Poul Johannes, 1979. Slipcast with a brushstroke of copper oxide under a clear glaze. Ht: 12 cm (4¾ in.), W: 26 cm (10¼ in.). *Photo: Poul Johannes.*

Porcelain lamp, Poul Johannes, 1984. Slipcast with engraved straw pattern and clear glaze. Ht: 17 cm (6¾ in.), W: 25 cm (10 in.). *Photo: Poul Johannes.*

Chapter 2

An artist's approach

What is it that draws a ceramic artist to include artificial light within their work? What are the possibilities? What fires their imagination, and what luck, skill and determination are also needed in order to complete the work?

There are many aspects to be considered by the creative maker, the first being the nature of the clay and the quality of light coming from within the form. Equally important is the effect the light surrounding it and falling onto the surface can have on the work. Other considerations include how and where the engineering needed for the electric lamp holder and the lamp will be housed and, most importantly, the space surrounding the form that the light will illuminate. It is for the artist to embrace all of these aspects and to create their own way of expressing their thoughts and feelings.

This chapter looks at a variety of artists, all working with translucent clay and light in interesting and diverse ways, showing a wide range of approaches to working with light and overcoming the related difficulties.

Mia Göransson

When I met Mia Göransson from Sweden at Collect in 2007, the first thing she said was, 'I talk with my hands.' Light is a vital ingredient in Mia's work. She works with different clays: 'The clay is what decides the feeling, and I catch things during my work by chance.'

I chose her sculptural works in bone china to illustrate the way she uses natural light as an integral part of her work. She has made moulds from leaves, casting them and then delicately attaching them to cast bone-china cubes. These she invisibly combines to make a wall installation in which they can resonate with the light in space.

Finding plain white bone china too cold, she developed ways of applying barely visible hints of coloured slip underneath the leaf that play tricks with your eyes as the faint tone reflects on the white fired-clay surface. She pours coloured slip onto the backs of the leaves before joining them, while still wet, to the square cubes. This formal repetition, and the simplicity of the concept, belies the depth of feeling within her work. Since these works were made in 2007, however, her work has changed greatly and brilliant colours now emphasize its inherent vitality.

Detail of wall piece, Mia Göransson, 2007. Cast bone-china leaf-form cubes, 60 x 60 cm (23½ x 23½ in.). *Photo: Margaret O'Rorke.*

ABOVE Mia Göransson's studio. *Photo: courtesy of the artist.*

BELOW Wall piece, Mia Göransson, 2007. Cast bone-china leaf-form cubes, 35 × 60 cm (13¾ × 23½ in.). *Photo: James Crabbe.*

Detail of wall installation illustrated on p.18, Mia Göransson, 2007. 35 × 60 cm (13¾ × 23½ in.). *Photo: James Crabbe.*

Kurokawa Kenkichi

Kurokawa Kenkichi, a Japanese ceramic artist, has a very different creative approach and experience. He enjoys the precision of cast forms. He creates sculptural models using earthenware clay and a plaster lathe. Plaster moulds are made, giving him hollow shapes that will contain the light source and retain the precise dimensions that interlock with each other. These repeated shapes are then high-fired. When they are assembled with the light installed, their crisp edges create expressions of light and shade.

Kenkichi uses a clay body called Han-Jiki, a type of china ware which is very white and resistant to distortions and collapsing. It is a widely used and not particularly special clay in Japan. He creates the slip by mixing the clay with water and silica soda, and then pours it into moulds. He works in a small studio crammed full of equipment for mould-making: a lathe, a plaster mixer, a slip mixer and an electric kiln. When I asked what he found most difficult in creating work using light, he replied, 'shrinkage and distortion, making the parts difficult to fit together'.

As well as being Professor of the Faculty of Design at Kyoto Seika University, and a freelance ceramic designer, Kenkichi is also a qualified electrician, so tackling the electrical part of his work is not so difficult for him.

ABOVE Kurakawa Kenkichi's work at an exhibition in Taiwan in 2007.

BELOW Wall lamp, Kurakawa Kenkichi, 2007. Ht: 25 cm (10 in.), W: 400 cm (13 ft). Han-Jiki. *Photos: James Crabbe.*

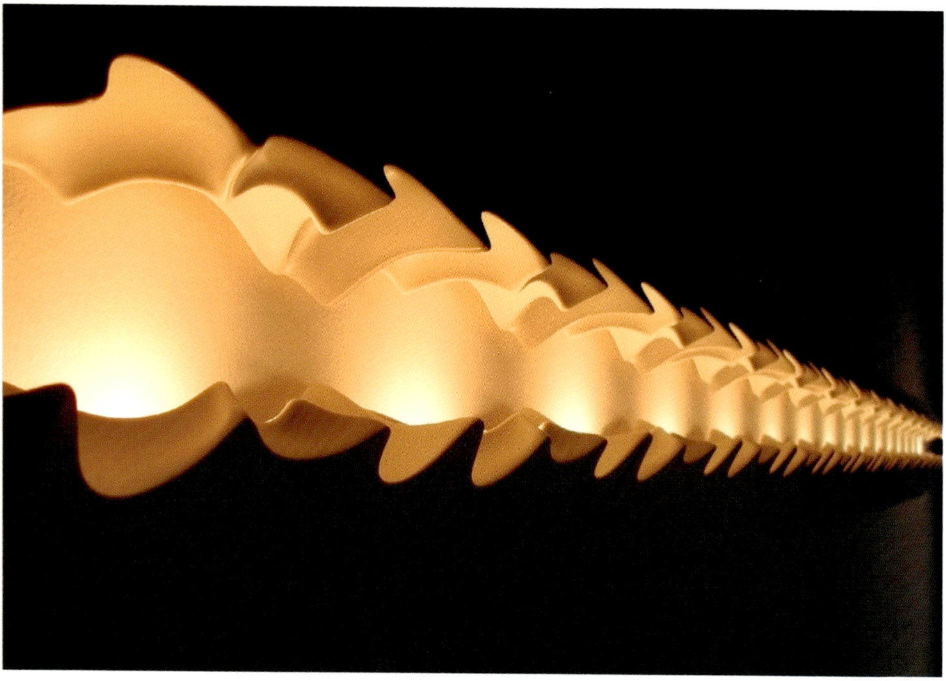

Wall Lamp (with vase), Kurakawa Kenkichi, 2007. Ht: 25 cm (10 in.), W: 400 cm (13 ft). Han-Jiki. *Photo: James Crabbe.*

Unitedlamp, Kurakawa Kenkichi. Ht: 12 cm (4¾ in.), W: 90 cm (35½ in.), D: 90 cm (35½ in.). Bone china and Han-Jiki. *Photos: James Crabbe.*

Unitedlamp, Kurakawa Kenkichi, 2003. Ht: 12 cm (4¾ in.).

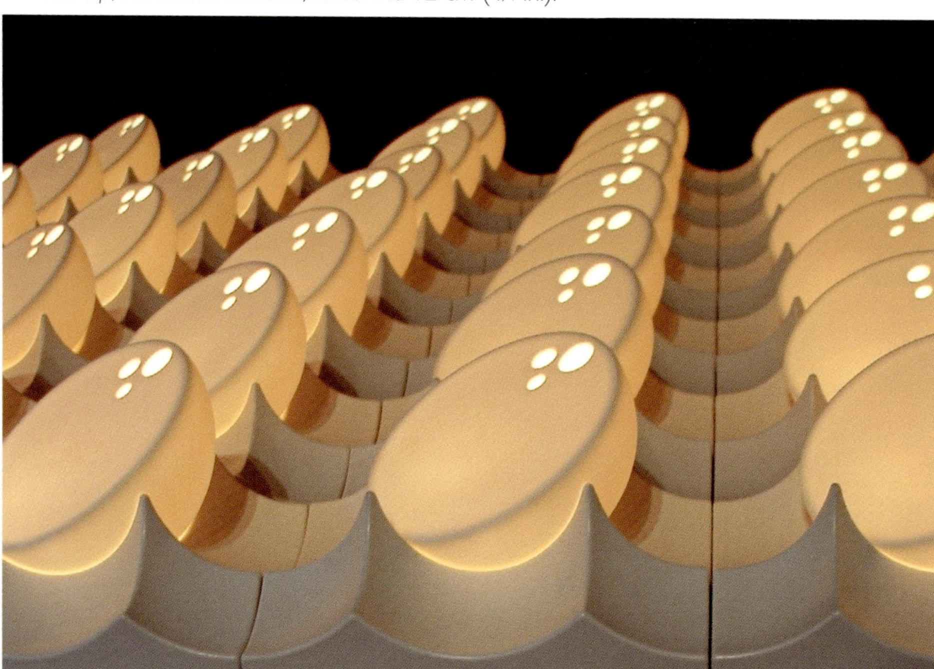

Kurakawa Kenkichi.

'The shape of the earthenware sculpture is first determined by new plaster moulds, and this is the most exciting moment in the entire procedure. Newly moulded pieces are extremely soft, moist and warm to the touch, like something living that has just been born. To complete the process the object is normally subjected to two rounds of firing, but this is the stage – when it has only just been removed from the mould – where we really sense its unique aura. This is the point where the artist feels a certain joy that no one else can, but it is also a time when he may sense acutely any deficiency in his skills.'

Margaret O'Rorke

A glazed ceramic vessel will reflect the light. A matt surface diffuses it. A translucent lit surface transmits light and can change both the atmosphere it inhabits and the way we feel. Recognising, questioning and understanding this is a basic consideration for a ceramic artist using artificial light within their work.

I share much of Mia's feeling about both the nature of the material and listening to the language from our hands. It took many years before I could find words to say what I felt about my work transmitting light. Ideas can come from anywhere at any time. A small notebook always kept with you is useful to record those fleeting moments. Some of these ideas from my notebook begin to grow for me as I become focused at the potter's wheel. I play with the material, enjoying its natural responsive movement in my hands, knowing it will, when high-fired, become translucent and be one of the strongest materials (except on impact). At the same time, subconsciously I am thinking about the translucency, visualizing the play of light and the space it inhabits, as well as devising ways to disguise the electrical fittings. I am interested in creating large-scale lit porcelain installations for

Kneading porcelain in my studio.
Photo: James Crabbe.

Thrown stack of wings, Margaret O'Rorke, 1999. Ht: 60 cm (24 in.). *Photo: Michael Harvey.*

Bubbles, Margaret O'Rorke, 1998. D: 50 cm (20 in.). Thrown porcelain. *Photo: John Woroniecki.*

Sea Frills, Margaret O'Rorke, 2002. D: 22 cm ($8^3/_4$ in.). Thrown porcelain. *Photo: Margaret O'Rorke.*

contemporary interiors. Porcelain, when lit, can completely change the atmosphere of a space, creating a calm and peaceful feeling.

When I first had the idea of using electric light within the porcelain almost 30 years ago, I struggled to throw tall thin forms high enough to contain the lamp holder and lamp (see photo in Chapter 1, p. 12).

As my skills grew so did my ambition to create larger-scale works. Spending three months during 1991 in the studio of Rioji Koie in Japan, where I was making my own work, heightened my awareness of how versatile the porcelain clay body can be. It was then that my adventures with clay really began, just as they still continue to this day.

Like other artists included in this book, it is my experience that finding the quality of light needed for each piece, and the type of lamp holder that will provide the light, then devising the structure to house them, are the most arduous parts. Porcelain is the most responsive clay to the touch, while to a potter the metal used in light fittings seems such an alien and unforgiving material. But the irrepressible urge to see the clay lit up drives me to overcome any difficulties in the course of devising and constructing installations for large-scale interiors.

LEFT *Vertical Pods*, Margaret O'Rorke, 2000. Thrown porcelain. Ht (each pod): 21 cm (8¼ in.), D: 20 cm (7¾ in.). *Photo: Gilles le Corre.*

RIGHT *Squirls Up*, Margaret O'Rorke, 2006. Ht: 30 cm (12 in.). *Photo: Margaret O'Rorke.*

Big Wheel (28 attached wings), Margaret O'Rorke, 2008. Thrown porcelain. D: 1 m (39 in.). *Photo: Richard Heeps.*

RIGHT *Wall Discs*, Margaret O'Rorke, 2002. Ht: 230 cm (90 in.), W: 112 cm (44 in.), D: 14 cm (5½ in.). Thrown porcelain. *Photo: Brent Ryberg.*

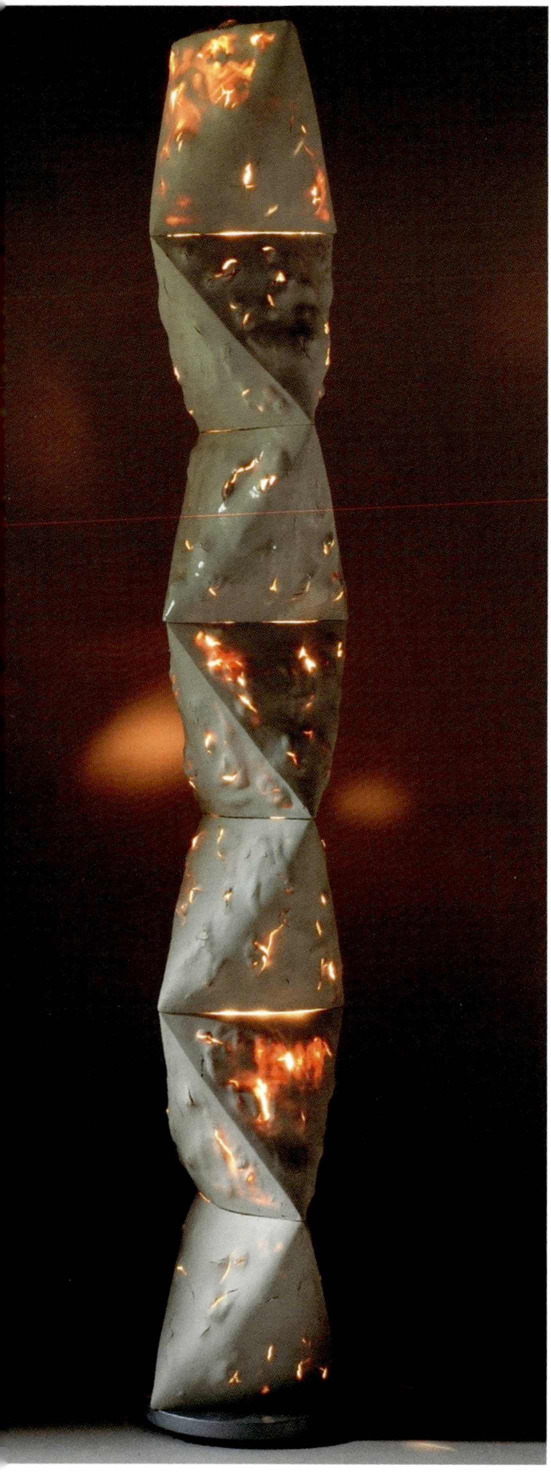

Tomasz Niedziólka

Tomasz Niedziólka is from Poland. He works with a wide variety of clays often dug straight from the ground. He is a lively thrower and frequently fires with wood. Tomasz's thrown natural clay works are imbued with vitality and freedom, and I was fascinated to learn that he had turned his hand to cast porcelain and electric light. This piece started as a cylinder, and was then re-formed. He used a knife from the inside to stab through the surface, allowing the light to escape and thereby freely create a surface treatment. This tall standing light is joined with two metal rods inside at the bottom and top. It's an ambitious work that seems to possess an extraterrestrial quality.

The sections used in this piece can be used in a variety of ways to create different constructions, they can be hung in one or more sections, or the column can be put together with more or less sections to create different heights.

LEFT *Column*, Tomasz Niedziólka. Ht: 170 cm (67 in.). Constructed from porcelain sections.

RIGHT *Column* (detail), Tomasz Niedziólka.

ABOVE *Supernatural*, Una Lee, 2006. Ht: 45 cm (18 in.), W: 360 cm (142 in.), D: 45 cm (18 in.). Cast porcelain installation.

BELOW *Supernatural*, Una Lee, 2006. Cast porcelain installation.

RIGHT Detail of *Supernatural*, Una Lee (as above). *Photos: Sh.*

Una Lee

Una Lee, a Gold Prize winner at the 4th World Ceramic Biennale, in Korea in 2007, is a Korean ceramic artist with a free spirit in her approach to porcelain and light. Her studio in Gangnam-gu (Seoul, Korea) is small, and space is limited. To my question, 'What inspired you to work with porcelain and light?', she replied, 'Light is necessary to view things. I am charmed by light as I perceive it through art and many things. I want people to feel a sense of emotion when they see my work. This is why I join light with my work. All of my work uses light, but there is more than one way to use light. I use it in a variety of methods.'

Una's work is very exciting. It has an infectious lightness of sprit, combining the clay with the light in a way that takes them into uncharted waters. She begins by using a computer as her graphic tool, after which she makes the clay forms as slender as possible. A plaster mould is made and the forms are cast. She uses an electric kiln and fires high, saying, 'The high temperature gives life and breath to my work. I like this method.' Because the scale of her works is ambitious (some 2 m/6½ ft plus), she often makes small pieces which, transformed by the high firing, can be assembled later.

Supernatural, from a series, Una Lee. 35 × 35 × 50 cm (13¾ × 13¾ × 20 in.). *Photo: Sh.*
RIGHT *Supernatural* from above.

Imagination or Reality 2, Una Lee, 2002. 4 X 2.1 m (13 x 6¾ ft). *Photo: Sh.*

When asked the question, 'What do you find most difficult?' her reply is interesting and illustrates the depth of feeling embedded in her work. 'All parts have difficult tasks. Electric wires, other materials, the combination of the two. I am always concerned about many things and try very hard to solve each problem. I especially worry about the quality of my work. Many of my works are joined together at exhibitions. My work is usually over two metres.'

Ian Johnston working in his studio.

Tea Light House, Ian Johnston, 1996. Porcelain, slab-built. 16.5 x 7.5 x 7.5 cm (6½ x 3 x 3 in.).

Ian Johnston

Ian Johnston is Canadian and works from his studio in Nelson, British Columbia. He is self-taught as a ceramicist and has a much simpler approach to combining light within his work, perhaps influenced by his previous training as an architect. He works with a variety of clays and forms, and first started to incorporate light in his work in 1996. When asked what inspired him to use light with porcelain he replied, 'The desire to push the material to its limits of thinness and durability. To express the fluidity of the medium with the aid of light.'

The stylized little *Tea Light House* has a jollity about it. Here he has played with the idea of using candlelight, avoiding the need to embark upon the complications of using electric light. The work is fired in an electric kiln. His studio is orderly and designed to follow his working pattern, beginning with the arrival of the prepared and bagged clay (purchased from Seattle Clay Products in the USA) and culminating in the packed departure of the finished work.

Jeremy Cole

Originally from Wellington, New Zealand, Jeremy Cole moved to London, England in 1999, and spent the next four years perfecting his design craft and production techniques. Strongly influenced by natural form Jeremy launched his design career with the introduction of his Aloe range in 2004, which was very well received.

As interest expanded, Jeremy set up his own studio in London to explore further the use of bone china with lighting, and

Blossom, Jeremy Cole, 2004. 80 × 75 cm (31½ × 29½ in.). Porcelain, acrylic, aluminium.
RIGHT *Orchid Cymbidium Lamp*, Jeremy Cole. Porcelain, glass, aluminium.

then moved to Stoke-on-Trent, in England.

Although he started out making everything himself, Jeremy now mainly designs, and all of his designs and products are developed, produced and shipped from his studio, where he employs a team of local ceramicists with 160 years experience in the ceramic industry between them.

Jeremy's lighting is inspired by the flora of New Zealand, and is painstakingly formed and crafted in bone china with each ceramic component being hand-finished. The translucency of the white bone china creates a beautiful glow when lit, and a totally different lighting effect is created through the production of the lights in different finishes.

Paula Bastiaansen

Paula Bastiaansen lives and works in Holland. The whiteness and surface texture of bone china inspired Paula to work with it. She is well-known for her extraordinary fine bone-china sculptures, and sources her bone china clay from Potclays in Stoke-on-Trent in the UK. Natural light and translucency have always been integral to her work, adding an ethereal quality.

Describing her studio she says, 'I have a very nice studio. It is a 200-year-old barn. On the outside I didn't change much, but inside I made a glass wall, so that there are two spaces. In one space I have a very comfortable studio, with heating and all the comfort that is necessary for a good studio. On the other side there is a place for exhibiting my objects. When I am at work, I have a beautiful view of the garden. I am very happy with this studio, it is a source of inspiration for me to work in such a nice place.'

Paula began making small objects in 2005, especially to incorporate

Light installation, Paula Bastiaansen, 2007. Bone china. *Photo: Marja van Hassel.*

candlelight, so that people who did not have the money for her large-scale complex works were able to own something simple that was made by her. The cluster of small, electrically powered lights illustrated here were made for an exhibition in Tegelen, Holland, where candles were not appropriate (for safety reasons). To make these forms she cut very thin slices of bone china 10 cm (4 in.) square, then wrapped them around a mould, and a minute later removed the mould while the clay was firm enough to handle but before it shrank around the mould too much.

Paula has ideas for larger-scale works incorporating light, but, like so many other ceramic artists, finds her greatest challenge is the need for assistance with the lighting and engineering.

Paula Bastiaansen in her studio.
Photo: Marja van Hassel.

Lilach and Ron Lotan

Lilach and Ron run the Lilach Lotan Pottery on Vancouver Island, British Columbia in Canada. I found their website on the internet, which was set up by Ron when he joined Lilach to run their pottery in 2005.

Lilach had been experimenting with thrown coral-reef forms on stoneware long before she came across translucent Laguna Frost Porcelain. This porcelain body fires to cone 6, and is a very good fine throwing plastic body. The work is fired in a 0.2 cu. m (7 cu. ft) Shimpo electric kiln.

It was the translucency that led Lilach to incorporate electric light. She has chosen to make her works fit ready-made IKEA lights in order to comply with US regulations, which require a

Lilach Lotan. *Photo: courtesy of the artist.*

Flowers, Lilach Lotan, 29 × 29 × 14 cm (11½ × 11½ × 5½ in.). Unglazed translucent porcelain thrown and joined.

Coral Reef Tubes, Lilach Lotan, 28 × 28 × 14 cm (11 × 11 × 5½ in.). Thrown and joined unglazed porcelain. *Photos: Ron Lotan.*

UL listing (Underwriters Laboratories lighting regulations). She has found this source to be both reliable and cost-effective.

Her studio is 100 sq. m (1100 sq. ft) and is divided into a main workspace, a glazing room, a kiln room, a packing room and a photography/storage room. See also p.87 and p.111 for more information and the website for the Lotans.

Inkeri Leivo

In the summer of 2008 I visited the Arabia factory in Helsinki in Finland, which in the past has provided a number of ceramic artists with their own studios on the ninth floor. These practitioners made their own individual works as well as designing commercial domestic ware to be produced in the factory. Now the artists in these studios have become a separate artists' cooperative. Inkeri Leivo has worked there since 1969. She is a prolific ceramicist and uses a wide range of different ceramic materials and techniques. She has received prestigious awards for the design of a number of best-selling domestic tablewares produced by Arabia, as well as creating a wide variety of her own individual studio pieces.

Inkeri began combining ceramics and light in the 1980s. The idea came from watching her mother, who worked on the fine rice porcelain ware in the factory. Inkeri wanted to create her own technique using light. In her own words, 'Clay is such a wonderful material; you can use it for making hard bricks or delicate, lucid objects.'

To begin with, the simple form is cast from a mould. She used the factory porcelain and created a method of her own, placing absorbent paper between the clay and the plaster mould, to depict individual forms that emphasize the translucency (while retaining the purity) of the clay. Each light is different. More recently, she has changed to using materials sourced from Arabia, including porcelain and bone china.

She says, 'Creating these lights demands a lot of patience. Usually the most demanding phase is the beginning when the clay meets the cast and starts to dry.'

Her work is fired using the Arabia factory's tunnel kilns and electrical kiln.

RIGHT Bone china lamp, Inkeri Leivo. Ht: 23 cm (9 in.). Slipcast. *Photo: Timo Kauppila.*

Bone china lamp, Inkeri Leivo, 1998. Ht: 23 cm (9 in.). *Photo: Timo Kauppila.*

Ilena Finocchi

I saw these lit bottles at NCECA 2007 in Louisville, Kentucky, in the USA. They are a literal representation of glass bottles, in which Ilena has portrayed the memories of her father, who had recently passed away. The ceramic bottles were cast from part of his bottle collection and the tools that were used to inscribe them came from his workshop. These objects are a poignant illustration of her memories of him, together with some of the skills he had handed down to her since childhood. To quote her reply to my question, 'What inspired you to work with porcelain?': 'Translucent porcelain caught my attention in graduate school, but it wasn't until the summer of 2006, during my residency at The Archie Bray Foundation, that I began to work with it. My dad passed away while I was doing the residency. The inspiration really came from looking at his life and his energy. Then I started to look at where our energy comes from and where it goes when it leaves our bodies. While looking for a metaphorical way to celebrate my dad's life and the memories I have of his energy, I also wanted to create a work that transcended the two of us to become more universal; light and translucency was part of that solution.'

Ilena was fortunate to have an uncle who is an electrician. He helped her to get started. She is now using energy-efficient lighting that has simplified the wiring, and she is also interested in developing ways of using natural light. When I asked her, 'What have you found most difficult in creating work using light?', she replied, 'Resources for different kinds of lighting and their limitations are my biggest obstacles. However, I have also found that

Life-size bottles, Ilena Finocchi, 2006. Cast porcelain.

obstacles require you to be more creative in coming up with solutions. It would be a dream to work with a manufacturer and design the lighting to fit into each of my custom pieces, but I am not at that point in my career.'

The photograph of the exhibition space shows clearly how thoroughly she has thought through the difficulties of installing and displaying her lit translucent works. She has built the stands for the bottles to look like a shelf, giving it a domestic feel and at the same time disguising the electrical flex for a temporary exhibition space in order to create an installation.

LEFT AND BELOW Bottles exhibited at the Archie Bray Foundation, Ilena Finocchi, 2006.

Liisa Rissanen

Liisa Rissanen is a young Finnish artist. I first saw her ceramic work in the international ceramic exhibition Born in Fire, held in Fiskars in Finland, in the summer of 2008. Drawn from a distance by the compelling simplicity of the raw-fired red-brick clay exterior, I peered inside and was instantly intrigued, asking myself, 'What is the story inside this tiny room of feet?'

This is Liisa's first work with light, inspired by one of her favourite poets, Raymond Carver. Her work has the same title as his poem *A Little Room*, about a relationship between a woman and a man who fight and then become silent. Light was already in her mind for this piece. She said, 'Light was there to enter and give shadows and space.'

The electrical lighting she engineered herself, using a simple low-energy screw lamp in a box beneath a sheet of frosted glass. For Liisa the greatest difficulty was deciding how many of the cast stoneware feet to use! The light showed all the defects. She thought of using black inside but feels now that white is the best option.

She plans to continue to develop ideas using light within an enclosed space, in contrast to the other artists represented here, whose work lights the space surrounding it.

Her studio in Fiskars is neatly organised, and she fires in an electric

BELOW Liisa Rissanen in her studio. *Photo: Margaret O'Rorke.*

The Little Room, Liisa Rissanen, 2007. Ht: 50 cm (19¾ in.), W: 36 cm (14 in.), L: 33 cm (13 in.). Stoneware. *Photo: James Crabbe.*

kiln to between 1200 and 1250°C (2192 and 2282°F). Liisa has added an interesting depth to the use of clay and light, and made it her own. I am looking forward to seeing where it will lead her.

The work illustrated in this chapter will, I hope, give the reader an insight into some of the varied possibilities that can be created with translucent clay and light. Most of the artists shown here have a website which will give you more information about their work (see p.111), and in Chapter 3, some of the artists' making processes are described in more detail.

Detail from the inside of *The Little Room*, Liisa Rissanen, 2007. *Photo by James Crabbe.*

Chapter 3

Making

'Let the process inform the work.'
For me, the word 'design' does not sit comfortably: it implies imposing ideas onto the material. Clay is the most abundant completely natural material, and is one of very few that can be fashioned directly by the touch of the hand. Creative ceramic artists need never lose touch with this fact, and will forever be able to celebrate it.

Perhaps it is because of my background and the time when I trained that I feel strongly about this. In the late 1950s at Camberwell School of Art, when Dick Kendall was Head of Ceramics and Lucie Rie first began teaching, design was not a part of our vocabulary, as far as I can recall. There were far more pressing issues for me. The clay, the materials, the firing process, the energy and vitality, our responses to the fine plastic qualities of porcelain, the feeling, the possibilities, and of course hard work.

Now things are very different. Computer-generated images can enable any form to be extruded if you know how the program works, and can relay the information to a machine that will generate a model. This is the recognized process for industrial production. Designers for ceramics no longer even have to touch the material. Like our hands, the computer is no more than a glorified tool, but does it have that immediate contact with all our senses?

For me, becoming a potter was a lifeline. Clay is a vehicle through which I can communicate feelings. It is direct, honest and exciting. As a studio potter, choosing to work with the translucency of porcelain raises many issues. All the processes become visible, and then, with the need to incorporate lamp holders and electrical wiring, the difficulties are multiplied. However, it can be done, as we have seen in Chapter 2. Artists are renowned for being ingenious at solving problems in order to materialise their creative ideas.

In this chapter I have concentrated on working with the clay. Just as a vessel maker learns to judge the volume of the container they are making, so the ceramic light maker needs to know how to realise the translucency of the material together with an awareness of the space the light will inhabit. There are no short cuts. In the electrical chapter, the technical and engineering aspects of the work are addressed in more detail, but you will find that both must be thought through simultaneously.

To begin with, the maker must learn about the material. Porcelain clay and bone china clay, although often grouped together, are totally different materials to work with, producing very different work when fired. Just squeezing in your hand a lump of soft clay from each clay body will immediately give you a sense of what might be made with it. Playing with forms and firing them at different temperatures will help you to discover how it reacts under heat, as well as its translucency when lit from the inside.

There are many different types of porcelain. Clay suppliers can be found on p.110. For most of my work in the UK I use Audrey Blackman Porcelain Body produced by Valentine Clays, which is currently exported into France, Sweden, Denmark, Finland, Northern & southern Ireland, Switzerland, Germany, Holland and Vietnam, and possibly the USA. This clay body was developed especially for its translucency by Dr Christopher Hogg at English China Clay. I have found it to be one of the best available porcelain bodies for throwing and translucency.

In the USA, Laguna Clays supply Laguna Frost Porcelain, recommended by Lilach Lotan for its translucency and for throwing, while Southern Ice, developed in Tasmania by Les Blakebrough for Clayworks, is also, I understand, very translucent. Laguna also supply translucent casting slips – recommended are NS125 Very White and NS131 Frost White.

Throwing with porcelain is not easy at first. It is a thirsty clay demanding a sufficient amount of water to allow it to move in your hands as the wheel turns; this in turn makes the clay wet and quick to collapse. To maintain the vitality of the throwing and the translucency when fired, it has to be thrown thin, which adds to the difficulty. However, with persistence you will find the balance needed and your forms will grow, just as a musician who diligently practises his instrument can learn to play wonderful music.

I make my tools from old clock springs and steel packing-case wire; this gives me the freedom to make a tool of any shape specifically for the job I want it to do (see the images below). Simply hold the metal with pliers and heat it to red-hot with a blowtorch where you

Tools made from clock springs and packing-case wire.

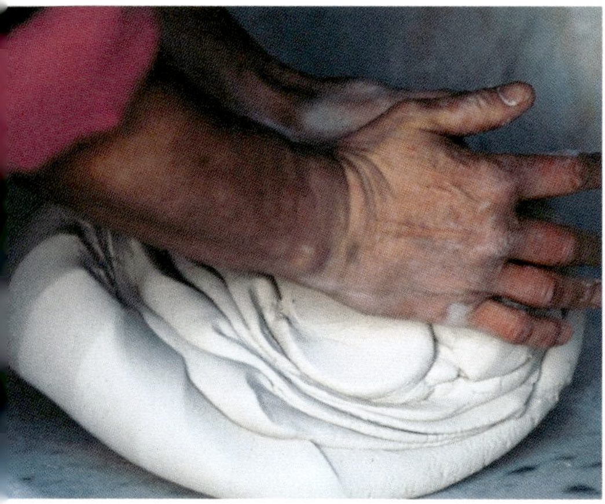

LEFT Margaret preparing the porcelain for throwing by kneading.

BELOW LEFT A tool made from clock spring, used for fluting.

BELOW Vessel, thrown, re-formed, fluted and fired.

FACING PAGE Free-standing porcelain light sculpture, Margaret O'Rorke, 1989. Ht: 33 cm (9 in.). *Photo: John Woroniecki.*

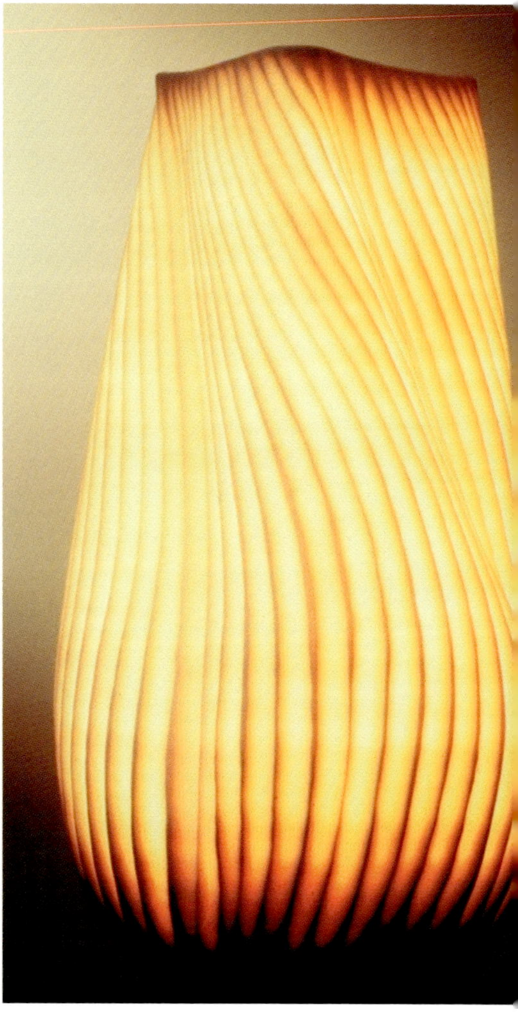

Kneading the baked rice biscuit into the clay. *Photo: Gilles Le Corre.*

want it to bend, then quickly immerse it in water to anneal the metal once it has been bent. Use string to bind it tightly to a wooden or cork handle. A metal file can give you a good cutting edge, but these tools will also sharpen themselves on the clay with use.

It is the 'what if' factor that leads to new discoveries. For example, thinking about the moon, I wanted to create a texture within the thrown form by allowing the process to create it. I baked some rice biscuits and, breaking them into pieces, kneaded them into the porcelain. As the form was thrown the texture naturally became part of the process. I will admit that I did expect the whole thing to blow up in the kiln when fired. Instead, with a slow burning-off period early on in the firing, all went well, and when fired it did look quite similar to a wonderfully full moon (see the photo opposite).

LEFT Throwing the form with the rice biscuit incorporated.

OPPOSITE The fired piece, by Margaret O'Rorke. The holes are created where the biscuit has burnt away in the firing. *Photo: John Woroniecki.*

This is another example in which the process has dictated the form. Thin wheel-thrown porcelain cylinders are cut in half and freely joined. Some ingenious engineering, devised to disguise the supporting structure and light fitting, has given the piece a life of its own.

LEFT Half cylinders, leatherhard and ready for joining.

BELOW *Thrown Half Cylinders*, wall piece, Margaret O'Rorke, 2007. From the exhibition Classic and Contemporary Pots, at Galerie Besson. *Photo: courtesy of Gallery Besson.*

The following caption sequence is a throwing demonstration, designed to give the reader an idea of the process involved in planning and throwing a large-scale work. The frame and electrical work needed for this piece are described in Chapter 5.

1 It all begins with a lump of porcelain.

2 Slapping it onto a marble slab to elongate it.

3 Then pinching it into shape.

4 The ends are cut at an angle to make a neat join.

5 The ring is secured on a dampened wooden bat.

6 Then thrown into a ring.

7 Then it is thrown and lifted into a cylinder.

8 A gas heat gun dries it enough to firm it up.

9 Another bat is placed on top to turn it over.

10 A gentle swing over turns it upside down.

11 This means that the bottom of the cylinder can be neatened.

12 A thin coil secures the bottom to the top bat.

13 The coil is pressed onto a dampened wooden bat.

14 The base of the cylinder is exposed and the bat centred.

15 The rim is neatened.

16 With the wheel turning, the heat gun firms the rim.

17 A sharp tap to the leatherhard cylinder forms an angle (see right).

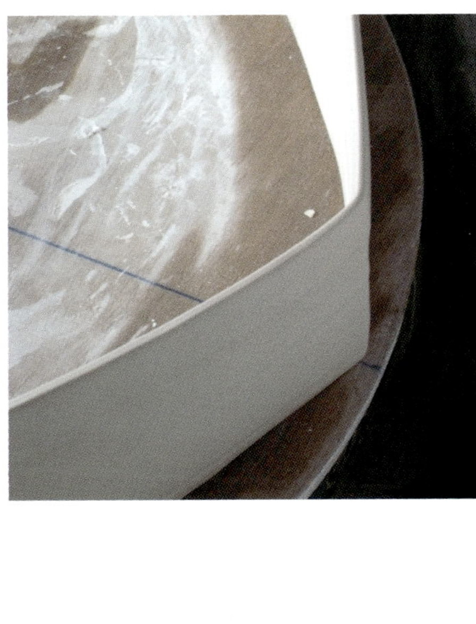

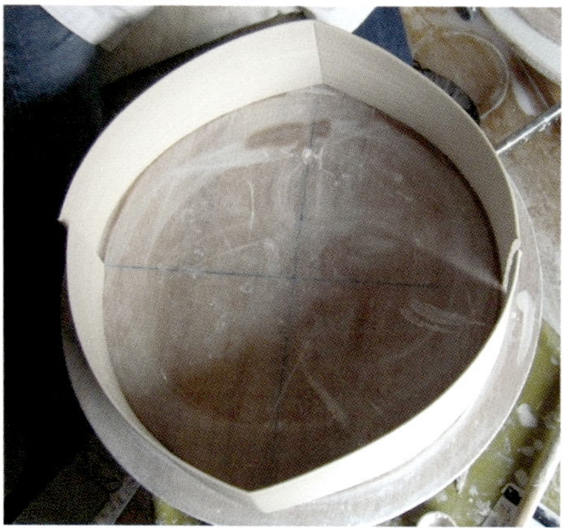

18 The rings are cut.

19 The edges are painted with vinegar, and then with slip.

20 The semicircles are brought together.

21 Carefully, they are assembled.

22 And then joined firmly, on the inside.

23 They are also joined on the outside.

24 Measuring the thrown ring so that each thrown tube is made to fit.

25 Throwing each tube to fit the height of the joined cylinders (see opposite page, above right).

26 The thrown tube.

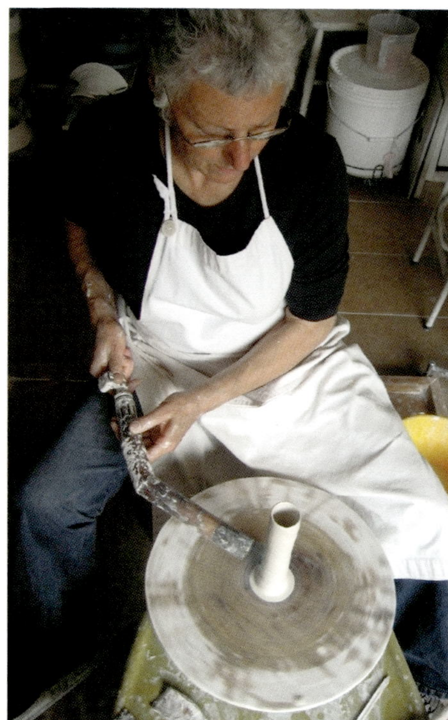

27 Firming it with the heat gun.

28 Vinegar and slip are added to both surfaces.

29 They are joined securely and the wing is complete.

30 Plastic helps to promote more even drying. Plumbers' insulating pipe helps slow down the rate of drying at the edges.

31 Drying continues, using wine bottles to help prop up the many sections.

32 Kiln props help to keep them upright in the firing.

33 Once fired they take on a life of their own.

34 Testing the idea on a temporary wooden stand.

The finished and assembled light. *Photos: James Crabbe.*

Bone china, although very translucent, is not a plastic clay and does not lend itself to throwing. It is more frequently used for casting and often needs support during firing.

However, Paula Bastiaansen has devised her own way of forming these bone-china light holders (see below), as mentioned in Chapter 2. Just as the process of making created the freedom within the forms in my wall piece shown at Galerie Besson (see this chapter, p.54), so Paula, by carefully forming the thin-cut square slices of bone china over the same mould, has produced a rhythmic, free-flowing pattern of light within this installation.

Sasha Wardell's cast bone-china wall light is an example of how the process can be used to produce a more controlled and formalised result. First she has cast with different coloured layers of slip, then, when it is dry, she has scraped away the layers of white bone china in certain places to reveal the coloured slip, creating a translucent lit pattern beneath the outer surface.

Liisa Rissanen has emphasized the process by modelling several legs, making the models in plaster and slip, casting lots of the same form for her piece *The Little Room* (see p.47).

Casting can create thinner, more translucent forms, but the clay is not always as stable as a wheel-thrown piece when high-fired. Solutions have to be found if you do not want to see the seams where the mould is joined when the piece is lit up, or where dribble has run as the slip was poured out. Also, bear in mind that the form may warp more readily. This can become an intrinsic part of the piece, and also depends on the type of porcelain slip that

Light Containers, Paula Bastiaansen, 2007. Bone china. Made for an exhibition in Tegelen in Holland.

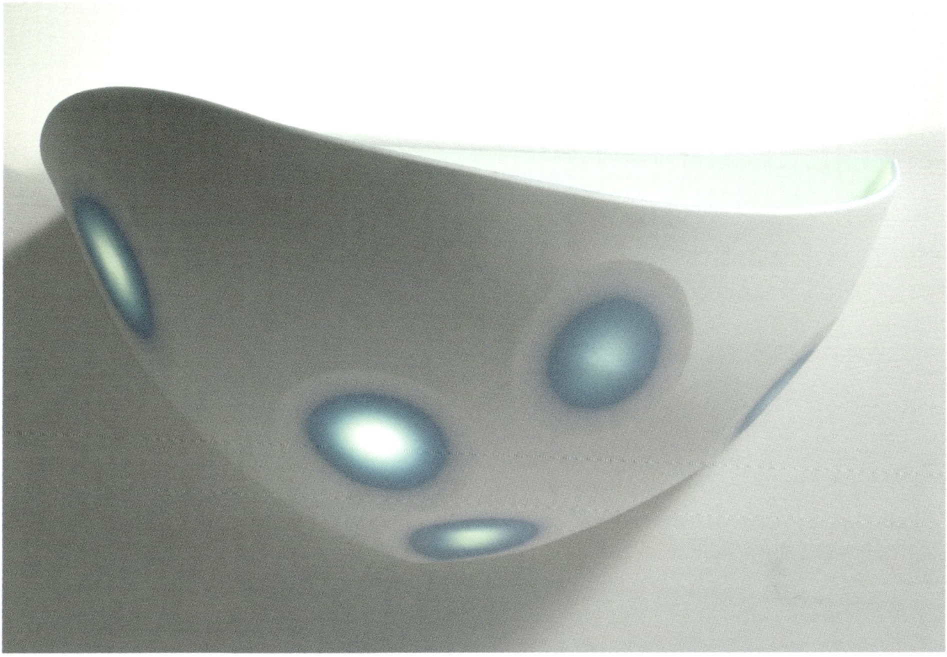

Space, Sasha Wardell, 2004. W: 38 cm (15 in.). Slipcast wall light, made from layers of coloured slip with carved decoration. *Photos: courtesy of Sasha Wardell.*

is used. Every ceramic artist will know there is only one way to find out: try it and see!

Several of the artists whose work is featured in Chapter 2 use cast bone china. This can be more translucent and is white or colder-looking when fired than many regular porcelain slips but does not lend itself to being thrown thinly. Instead, forms can be thrown thicker and then extensively turned when leatherhard, or else carved or worked on in other ways. This can, however, mask the vitality of the throwing and without care can easily lead to static forms.

Sasha Wardell uses layers of bone china slip, and then scrapes back layers to create areas of translucent colour.

Chapter 4

Working with Industry

A creative ceramic artist, working with the translucent nature of porcelain or bone-china clays, can produce lit translucent works that transcend the conventional lampshade-type structures and change the atmosphere of a space. What is needed is a sensitive understanding of the nature of the material and the complex ceramic process, together with the ability to visualise the work in space. Ingenuity is essential to devising ways of combining the material with the engineering needed to integrate the light fitting.

The particle structure of porcelain, and of bone china, together with the amount of oxygen present during the firing, gives life and warmth to ceramic forms when lit. A slipcast form made from a mould of a clay model illustrates the ceramic skills of the maker when the fired piece is lit. The advantages of ceramic lights are that the fired form does not need any supporting structure or electrical insulation (unlike a paper form). It is resistant to heat and is antistatic. The fixtures can be part of the form and remain unseen.

An artist in close touch with clay can bring new and exciting ideas to a wider audience by working together with technicians from the ceramics and lighting industries. However, there are fundamental differences between working as an independent studio artist and designing work for an industrial manufacturer.

As a freelance designer, it is not easy to find a lighting company to invest in your ideas. If you are new to the lighting scene and want to develop commercial porcelain lighting to be manufactured, it will help if you do your research first. Visit lighting shops and department stores. You will learn a lot about fixtures and fittings. See where there is a gap in the market. Try to get a feel for which area of the market you will be targeting. Look at current architecture. Attend appropriate design exhibitions and events to see what other contemporary designers are making. Most importantly, will you want to see a lot of them in ten years' time? Manufactured products can enable your work to reach a wider audience at an affordable price. We have a responsibility as artists to develop exciting, intelligent ideas that can be manufactured in quantity but will also stand the test of time.

Many of the skills and production methods are similar to those we use in our studios. First-hand experience in a ceramics factory is invaluable in enabling the artist to understand and adapt to the rigours of an industrial environment, namely, the detail required and the pressure imposed when designing a commercial product.

Once you have developed a prototype that you feel could be of interest, look for the lighting companies that most suit your ideas and approach their design departments. You have nothing to lose, and you'll gain experience if you are lucky enough to be offered an interview. Always remember that everyone's time is precious.

Wade's porcelain factory, Stoke-on-Trent

In 1999, after 19 years as a self-employed ceramic artist, I was awarded a research and development grant from Southern Arts. This enabled me to approach Wade's porcelain factory in Stoke-on-Trent. I took some examples of my work to an interview with the MD, who invited me to spend three months in the factory. By the time I left I had produced a prototype. I threw the model, and the mould-maker Cyril Roberts, who worked at the factory, taught me how to make moulds and how to slipcast them. One mould created three different products: one on its own formed a wall light, two together a freestanding piece, and three lying back to back formed a ceiling and/or large wall light (see below and p.71). Wade's tested the viability of the design and agreed that this porcelain form could be produced commercially. I felt ready to create a range of commercially

Cyril Roberts making a block from the cast light mould I had made. *Photo: Margaret O'Rorke.*

viable porcelain lights and collaborate with a lighting company. (The engineering and electrical design work that followed is covered in Chapter 5.)

These three different products made from one mould reduced the production costs, and gave a cohesive design

Cast porcelain wall light, Margaret O'Rorke. W: 43 cm (17 in.), Ht: 18 cm (7 in.). *Photo: Richard Heeps.*

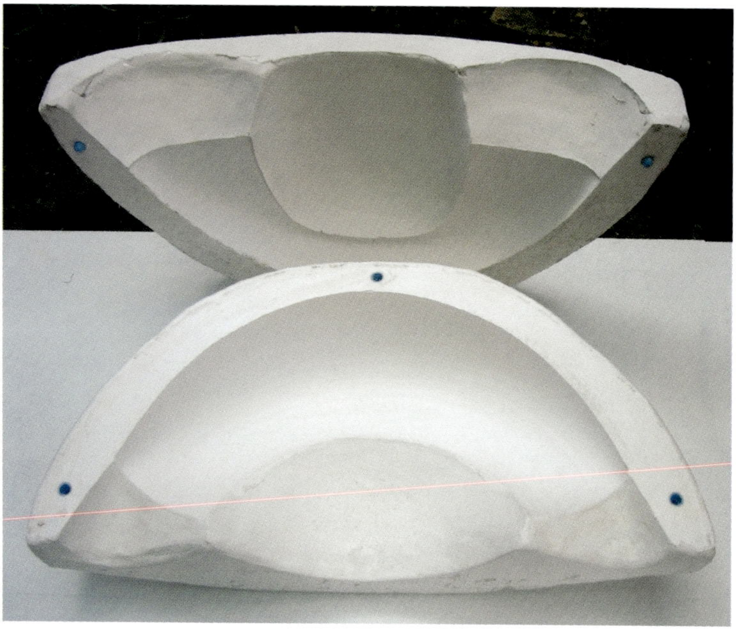

The single plaster mould from which all three lights were cast and adapted.

package for marketing purposes; identifying your market and finding an appropriate distributor willing to make a financial investment to cover the production, marketing and sales costs is a key issue. I currently cast and fire the forms in my studio.

Tragically, like so many other ceramics factories, Wade's has now closed, with the loss of many jobs amongst a highly professional, skilled and dedicated workforce. However, the knowledge I gained at Wade's gave me the skills to take on my first large-scale commission for a cast chandelier, which was produced in my studio but could be adapted for commercial manufacture. The commission was based on a free interpretation inspired by a wisteria branch. VBK Lighting designed the metal structure.

The aluminium frame I designed to hold the mould and to tip out the casting slip. *Photos: James Crabbe.*

ABOVE Cast porcelain ceiling wall light, Margaret O'Rorke. Adapted from the same concept. Dia: 43 cm (17 in.). *Photo: Richard Heeps.*

BELOW Free-standing cast porcelain light sculpture on a slate base. Ht: 23 cm (9 in.), L: 43 cm (17 in.), D: 16 cm (6¼ in.). *Photo: Colin Hall.*

Wisteria chandelier

Wisteria in bloom.
Photo: Margaret O'Rorke.

1 Thrown model.

2 Three sizes of biscuit-fired thrown models. *Photos: Irina Tatarnikova.*

3 The plaster mould is made from the thrown biscuit-fired model.

4 The finished plaster mould.
Photos: Margaret O'Rorke.

5 Pouring porcelain slip into the mould.

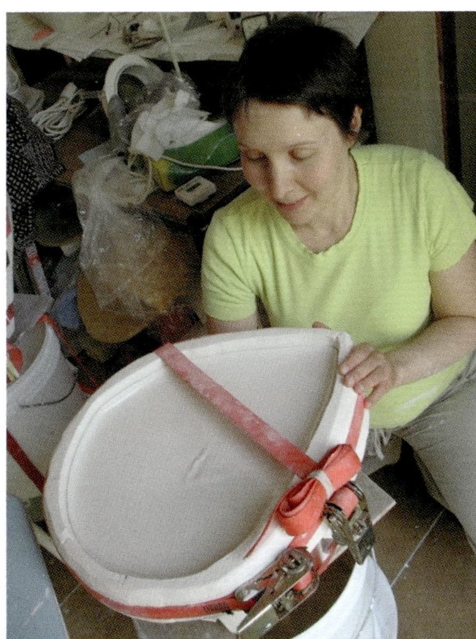

6 My assistant Katja Bjerrum releases the slip through a stopper at the base.
Photos: Irina Tatarnikova.

7 The slipcast form is cut and re-formed in the mould while still soft and plastic.

8 The re-formed piece is released from the mould. *Photos: Irina Tatarnikova.*

9 The shape is placed in a pre-thrown chuck.

10 It is then covered with fine plastic for slow drying.

11 Some of the 55 blooms in their thrown chucks ready to fire.

12 The buds in the foreground.
Photos: Margaret O'Rorke.

13 The kiln packed with the wisteria pieces. *Photo: Margaret O'Rorke.*

14 Metal bracket attached.

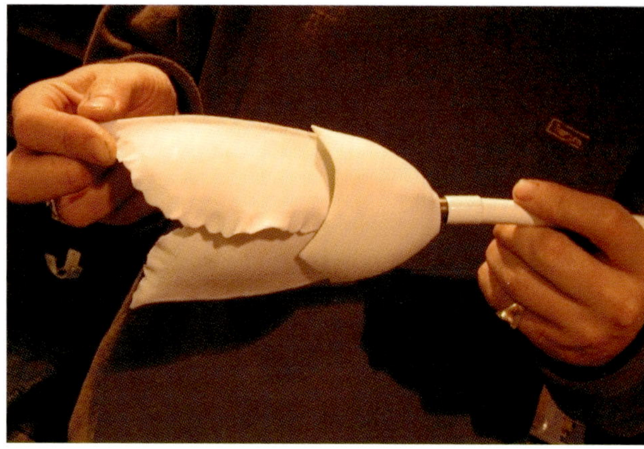

15 Testing size and fitting of the bud cup.
Photos: Margaret O'Rorke.

16 Testing the fittings in the workshop. Metal frame designed by Max von Barnholt.

17 The blooms lit up.

18 Installing the chandelier on site.
Photos: Max von Barnholt.

Wisteria chandelier, 4 m (13 ft) high, installed in the stairwell of a chateau in France.
Photo: Antjy Geczy.

The Royal Copenhagen Porcelain Factory.

RIGHT The vase I chose to work from.
Photos: James Crabbe.

Royal Copenhagen porcelain, Denmark

A later industrial encounter, in 2006, gave me the opportunity to work in the Royal Copenhagen porcelain factory, which was newly relocated to a brand-new state-of-the-art factory built outside Copenhagen.

The design brief was to fill a 9-metre (29 ft) high empty void. I was to use one of their existing vase moulds to create a permanent porcelain lighting installation for their renovated showroom in Amagertorv in the centre of Copenhagen.

Their Royal Copenhagen porcelain body is fired to 1400°C (2552°F), at which temperature the form is very soft. I employed some of the techniques I had developed with my thrown studio pieces and devised a way to re-form the vase shape, taking advantage of Royal Copenhagen's slip, which is very translucent when thinly cast.

Two highly skilled craftsmen, Bent Nødskov Jensen (Bo) and his colleague Thomas, who manually cast the large moulds, gave me such unwavering support that we made 60 lights in 10 days (see p.84).

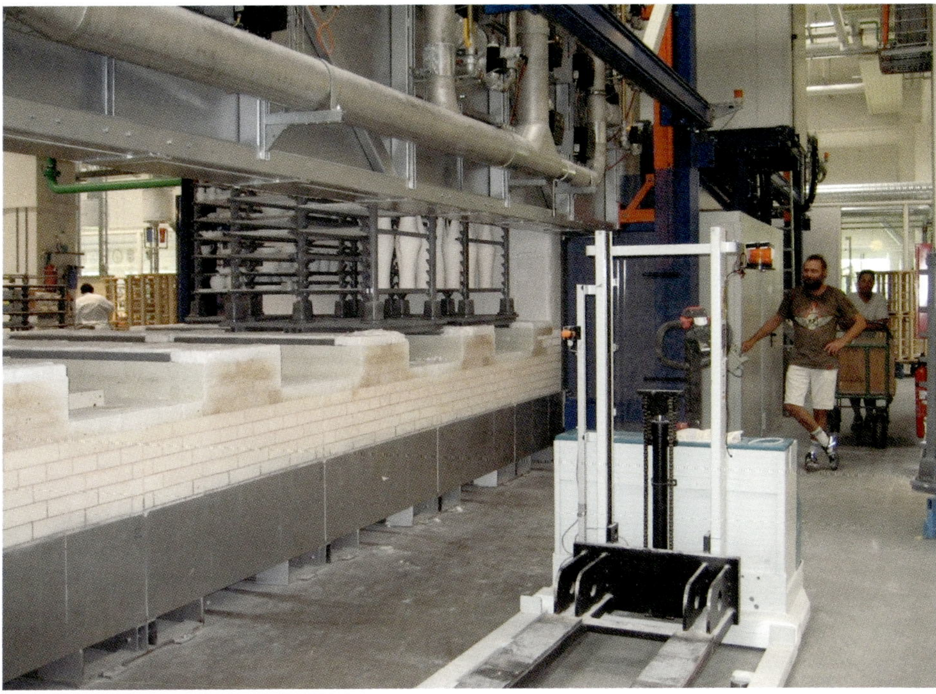

The kiln. The doors lift up both sides at the press of a button. *Photo: Margaret O'Rorke.*

Me, Bo and Thomas experimenting. *Photo: James Crabbe.*

The light we made. Height: 43–44 cm (about 17 in.). *Photo: James Crabbe.*

ABOVE AND BELOW Bo and Thomas casting the lights from the existing vase mould. *Photos: Margaret O'Rorke.*

1 Cutting the sides.

2 Adding slip to the edges.

3 Joining one of three sections.
Photos: James Crabbe.

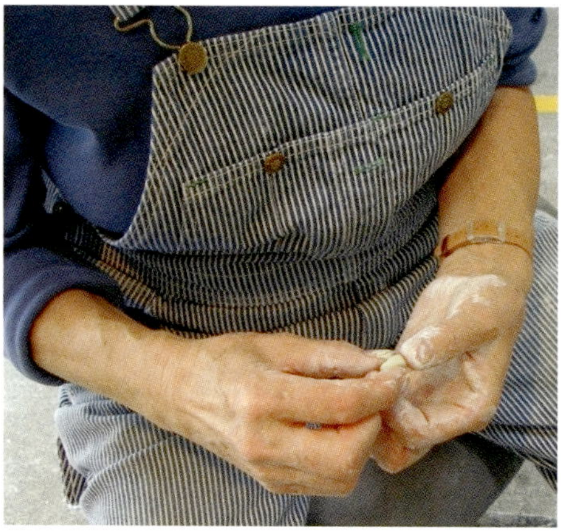

4 Making the lug to hold the fitting.

5 Making the lugs.

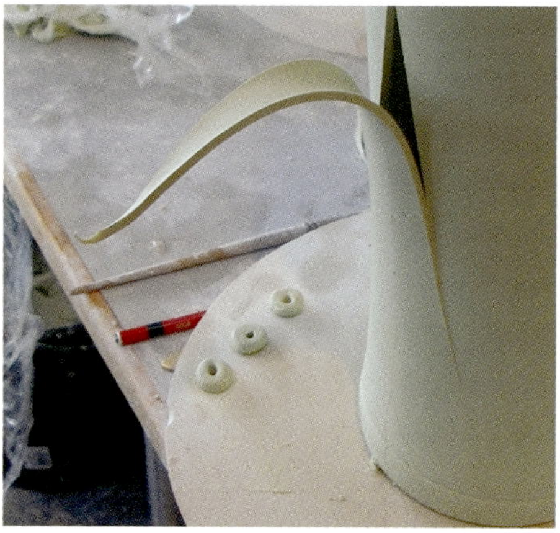

6 The finished lugs.

7 Placing the leatherhard slipcast form into the chuck, to support the form and pointed ends at high temperature. *Photo: Bent Nødskov Jensen.*

8 Cutting the top.
Photo: Bent Nødskov Jensen.

9 The first batch ready for firing sitting in their chucks (for protection).

10 The first fired piece out of the chuck.

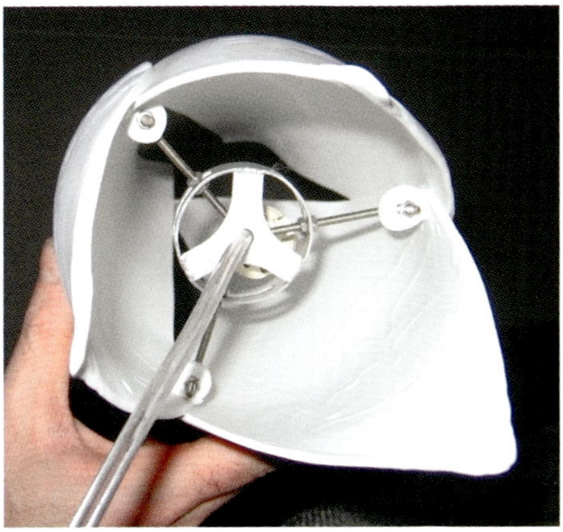

11 The stainless-steel bracket is fixed in place, into the lamp holder, designed by Jesper With. *Photos: James Crabbe.*

12 The first row installed at Amagertov. *Photos: James Crabbe.*

13 View from the balcony. *Photo: Margaret O'Rorke.*

The installation at the Royal Copenhagen showroom on Amagertorv, in the centre of Copenhagen. Made at their porcelain factory in 2006. View from below. *Photo: Margaret O'Rorke.*

Approaching manufacturers

Having a good idea and design is the most important thing, but once you feel confident with your creation, and would perhaps like to try to work with industry, these are some things to consider when approaching factories. It is a complex process to combine a knowledge of the material, understand the engineering needed to make it work, and develop a marketable design. To add to that, you need to learn how the ceramic and electrical manufacturing industries work, be able to work with their staff, and be prepared to make compromises where needed in order to produce a marketable product for people to enjoy.

The artist is the link between the ceramics factory and the lighting factory, and as such needs to understand the technical requirements for the lighting in order to create the piece as a whole. First you will need to find a lighting company. They will be funding the research, manufacturing costs and marketing. They will most likely have identified their own market and will only be interested in a product that fits within their chosen range.

When you know the name of the company you wish to approach, find out the name of the design director, and send a short letter with a few images to grab their attention and give them an idea of your work. If you are invited to an interview, listen carefully to what their policy for design is first, and if you feel you can work within their brief then discuss your ideas. Do you think you will have the right skills to enhance their business? They cannot afford to take risks, so you need to be clear on your project.

By this stage you will already have acquired the skills needed to understand the mould-making process. The simplest solution is often the most successful, but also the hardest to find. If you have made a prototype then take it with you, as this will be key, and set it up as fast and as neatly as possible.

After spendng two months working at Wade's, the marketing director came into my studio and said, 'I want to see your product!'. Looking around, I was not sure what he meant as I had never thought about my work in this way. I already had one cast wall form lit up that he could see, and he explained that he wanted a set, much like a 'hat, handbag and shoes'. We discussed it and I agreed to make two other forms using the same mould (see p.71). Producing sets is a marketable approach. If your designs are aimed at the high end of the market you could consider a limited edition.

Timing is fairly crucial for production and marketing, and it usually takes two years from concept to finished product. There are many stages to go through. If your project is taken up, a contract will need to be drawn up, so make sure you understand what you are signing. There is help available – AIR (Artists' Interaction and represenattion) can be found at www.a-n.co.uk/air and can advise you.

A lighting company may work directly with one ceramic factory, but if not you will need to find a suitable company used to working with your clay. In the UK, Stoke-on-Trent is the first place to look. Many companies have recently closed due to competition from cheaper labour in the Far East, but many factories are still there, and by supporting them we can help to keep the manufacturing culture alive.

Finally, keep an open mind. You may need to make compromises, but always maintain the highest standards.

ABOVE AND BELOW *Kou Ceramic Chandelier*, Mark and Emma Welsh. Made in three sizes: 63.5 cm (25 in.), 88.5 cm (34¾ in.) and 140 cm (55 in.). *Photos: Mark Welsh.*

In contrast to this, you may be approached to discuss a commission, so listen to their brief first, and have with you a list of any important questions – make sure you take down notes.

Mark and Emma Welsh are a young partnership living and working close to Stoke-on-Trent. They wanted to help revive the rapidly diminishing industry in the area by bringing in fresh ideas with contemporary ceramic designs (see images opposite)that could be manufactured by the highly skilled workforce still remaining in the area . Their business has developed rapidly through Mark's skilful use of the internet and Emma's intelligent approach to producing commercial contemporary lighting designs. Emma, like many other artists working with clay and electric light, finds it challenging to develop designs that can be manufactured. The ideas begin in her studio, then with the help of drawings she discusses the making of the model and moulds with an expert mould-maker. She uses a professional electrical engineer for the lighting technology.

Mark has a background in marketing and computing, and has created an extremely clear and comprehensive website (listed on p.111) giving widespread promotion to their products and attracts worldwide sales.

Theirs is an excellent example of collaboration between two skilled people who have been able to develop for themselves a successful industry of their own, enlisting the skills of the experienced workforce around Stoke-on-Trent. Discussing the promotion and marketing of their work, Mark says 'the costs of developing a website and buying exhibition space are roughly comparable ... but the longevity of the investment's return makes a website excellent value.'

Lilach and Ron Lotan, whose thrown porcelain lights are included in Chapter 2 (see pp.39–41), are another successful young partnership, based in Vancouver. With the combination of Lilach's ceramics and Lotan's management and computing skills they have promoted their business on the internet with a comprehensive website (see p.111), and gained an international clientele.

Jeremy Cole also promotes his individual, finely crafted lighting designs from a website as well as from retail outlets. Jeremy launched his design career with the *Aloe* range (see p.37). *Aloe* is made from individual cast porcelain leaves that are geometrically arranged around the light source to act as a diffuser, producing a calm, glowing light. He has set up his own workshop in Stoke-on-Trent to manufacture his products (see his website listed on p.111). All of his designs and products are exclusively developed, produced and shipped from his own manufacturing unit, where he employs a team of local ceramicists.

The future

There is a lot of unexplored potential for commercially manufactured translucent ceramic lighting products and installations. What is needed is collaboration between the artist (who has a good understanding of the nature of the material and an imaginative and adventurous approach), a skilled ceramics workforce and a lighting industrialist willing to promote and sell the work. Architects and interior designers able to imagine the untapped possibilities of the material have a very important part to play in helping this to come about.

Chapter 5

Electrical

This chapter introduces some of the practical implications of incorporating electrical lighting within the ceramic artist's translucent porcelain or bone-china art works. It will also help with preparations towards making a model prototype for industrial design, prior to approaching a lighting manufacturing company. If you have no understanding of basic electrical wiring regulations and safety precautions, you need to acquire it before beginning to wire up lamp holders and plugs. This chapter aims to introduce the reader to the essentials. In this regard, there are a number of helpful websites identified on p.97.

The artists I contacted for this book unanimously agreed that finding the ideal electrical light source for their work and incorporating the lamp holder and lamp within the work is the most difficult aspect.

I have identified some useful types of lamp holders and lamps to enable the artist to begin to discover how the light fitting can work within the translucent porcelain or bone china and transform the space surrounding it. However, you must consult the safety regulations for the country in which the work is to be sold before you can consider selling your work.

As I mentioned in Chapter 4, it is helpful to begin by looking in retail lighting shops. Study the technology of the lamp holder: what is it designed to do, how is it incorporated, what type of lamp holder and lamps are used, and what is the light produced actually doing? Also, look at the different types of flex used for the free-standing pieces. This research helps to focus thoughts on the function of the work and the space it inhabits.

A ceramic artist will need a four-dimensional imagination in relation to space and function, as on top of the 3D form you will need to consider the translucency of the clay and the quality of light emitted. The form and the technical solution, along with the function must all be thought about together. Aspects to consider include the nature of the material, translucency, the firing process, electrical fittings and the light to be given out.

In the early stages, while exploring your ideas, rough sketches will help to clarify some of the practical issues you will encounter. Making a three-dimensional model can also help to develop your ideas and identify practical issues. It is no easy undertaking for a ceramic artist to maintain a creative approach to clay, while combining the complex metal engineering and electrical work needed.

Careful research and thought must be engaged in to find the ideal light source. What do you want the light to do? Which would be the most suitable light source? How can it be incorporated?

It is important to think the whole thing through before you begin making.

Sourcing the lamp holder needs to be addressed early on so that you can begin to see the effect the light has on the ceramic. This will help you to work out what form the fixing plate with the lamp holder must take, and how the ceramic will be attached and removed from it so that the lamp can be changed and the work cleaned. The same basic principles apply to those using computer-generated design programs.

While considering these practical issues, remember that the underlying reason for combining the translucent porcelain or bone china with light is the nature of the material, the forms that can be created and the quality of light that only the translucent clay can transmit. The ceramic needs no support when fired, unlike fabric or paper; is anti-static and non-conductive, so does not attract dirt; and doesn't conduct electricity.

It is the work of the artist to experiment with the fired ceramic to find out how the light will look in conjunction with the translucent material. Using your knowledge of the nature of the material and the firing process, it is possible to begin to imagine in the mind's eye how the light and the clay will interact.

A stable standing drill. *Photo: James Crabbe.*

To reach this stage involves some basic electrical knowledge and also some practical engineering skills. Now is the time when the difficulties of maintaining creative freedom really begin to become apparent.

It is difficult to think through all the implications for the material, how much it will re-form and shrink when high-

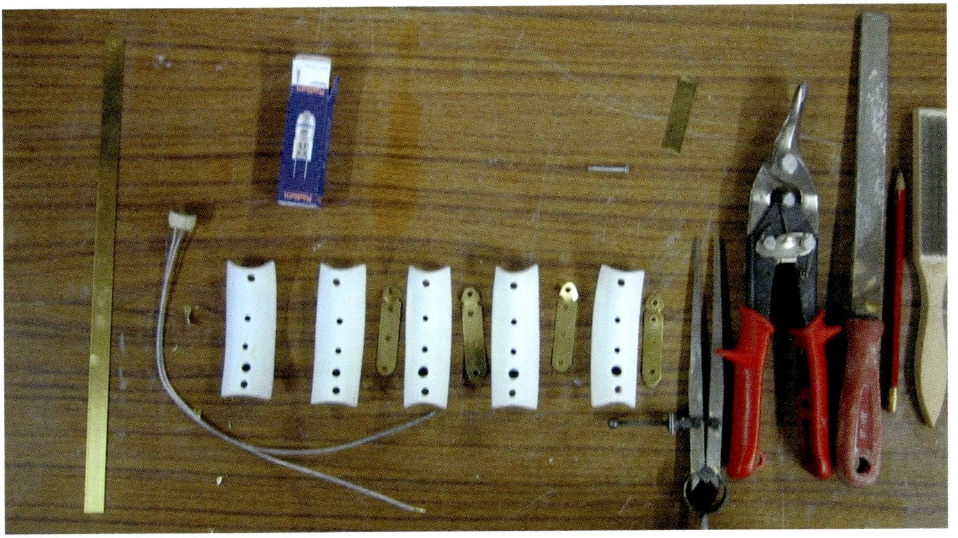

Examples of tools and preparation for making fixings and brackets. *Photo: James Crabbe.*

Vertical Pods by Margaret O'Rorke. Thrown porcelain light sculpture wired with a 12-volt transformer and a 20-watt halogen lamp. *Photo: Gilles le Corre.*

fired, at the same time as considering the practicalities needed to install the lamp holder and fitting.

When developing and creating new forms, always have good-quality tools for metal and electrical work. These should include: a stable workbench and drill, a hacksaw, wire cutters, a screwdriver, pliers, G-clamps, a wood saw, a jigsaw, and a stable drill with high-speed drill bits. This set-up will give you the freedom to construct the support, and to experiment so as to achieve the effect you want from the light source exactly where you want it.

The *Yellow Pages* and the internet can help you find a local metalworking company for sheet metal, rods, etc. They will be able either to supply you with metal offcuts for your prototypes, or to direct you to someone who can.

If your ideas need a more substantial steel structure, a metal fabrication company will be able to help you, using heavy machinery for cutting and bending straight-edged forms and a laser cutter for curves.

If you are cutting the metal yourself, aluminium costs more but is much easier to work with. Thin sheets of brass or aluminium for making small fixings can be found in model shops.

M3 nuts and bolts are a useful size for joining metal to metal and are standard for connecting some lamp holders to metal wall and ceiling fixtures. Pop rivets, available from any DIY shop, can also be useful to join metal to metal.

When the metal fitting is finished it can be painted. Rub it down with fine wet & dry paper, then use a spray can of engine-components cleaner to remove any grease and loose particles, before applying a coat of car spray primer, then two coats of car spray paint. All are available from motor accessory stores.

Electrical safety is paramount. Voltage supply is either 110 volts in the USA/Canada or 230–240 volts in the UK, the EU and most other countries. Check legal regulations for electrical lighting in the country where they will be used. Always check the voltage supply and electrical plug before beginning your work. It is also advisable to check the UL safety standards (www.ul.com) in each country before selling your work (see p.96).

Take intelligent safety precautions to ensure both your own safety and that of the people around you. Remember to disconnect the plug from the mains or the fuse box before touching the lamp holder. Never take short cuts with electrical wiring. Seek advice from professionals if in any doubt and before selling to a client.

Finding the right lamp holder and lamp is not easy. A well-informed electrical-lighting technologist can help steer you through the vast range. In my own case, Beverley Bigham of Preferred Electrical UK has been invaluable in helping me to find the type of light source, lamp holder and flex I need for my work (see the list of suppliers at the end of this chapter for contact details).

The standard bayonet fitting used in the UK, as well as the screw-fitting lamp holder used in most other countries, is readily available. These fittings house incandescent lamps, which give electric light by heating a tungsten wire filament. The lamp holder is wired directly to the mains plug. The flex must be restrained by the fitting before leaving the lamp holder so that the live wires can never become exposed.

A laser cutting a thicker steel base.

Incandescent lamps generate considerable heat and are rapidly being replaced by a wide variety of low-energy compact fluorescent lamps that fit the same standard lamp holders. Make sure the lamps are the correct voltage for the country of use.

There is a variety of decorative two-core and three-core lighting flexes available that must be double-insulated.

Threaded rod, basic screw and bayonet lamp holders, plus nuts, washers, bolts and flex restrainers needed for this standard lamp holder, as well as the Halopin lamp holder illustrated below can be obtained from Preferred Lighting and other companies (see list of useful websites on p.110).

For small forms that are lit I have found two types of lamp holder useful. Firstly, the low-voltage halogen GY6.35 12-volt lamp holder with leads (see image on p.84) can be wired directly to the lamp holder, or is connected to a transformer using small-size connector blocks. The

GU5.3 lamp holder 50 mm (2 in.) diameter dichroic reflector spot lamps.

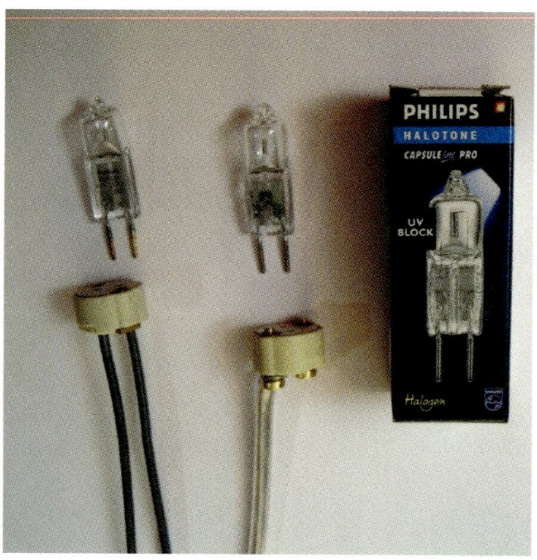

GY6.35 with M3 threaded bushes.

safest and most convenient lighting transformers are those incorporated in the plug. Make sure you choose the correct wattage for the lamps to be used, and check the voltage of the country in which it will be used.

The GY6.35 12-volt halogen lamp is available as either frosted or clear from 20 to 100 watts; 50 mm (2 in.) diameter dichroic reflector spot lamps from 20 to 75 watts are also available and fit the GU5.3 lamp holder. These lamps have a 2000–5000 hour lamp life depending on quality. The Halopin screw fitting is versatile because it is wired directly to the electrical mains supply and does not need a transformer. They give the equivalent of a 20–75 watt light and are a small and compact size.

The G9 lamp holder does not need a transformer and is wired directly to the mains. The flex must be double-insulated. If the wires are multi-stranded, use a soldering iron to provide a solid end to push into the fitting. Heat the wire first then engage a small amount of solder. Locate the small 'L and N' holes (live and neutral) on the bottom of the lamp holder. Two of each hole are available, so that these lamp holders can be linked to

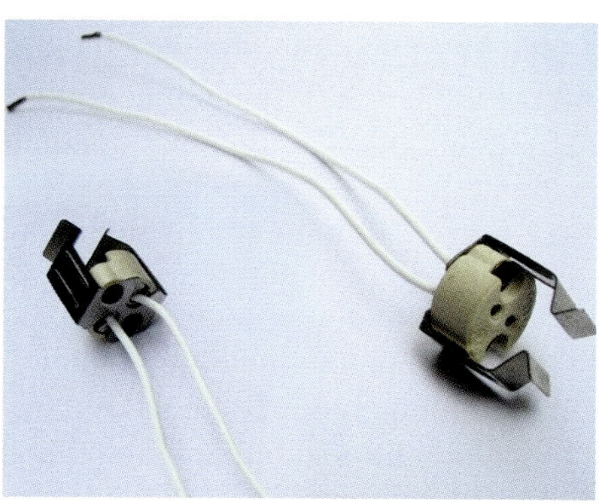

GY6.35 with leads and M3 threaded bush.

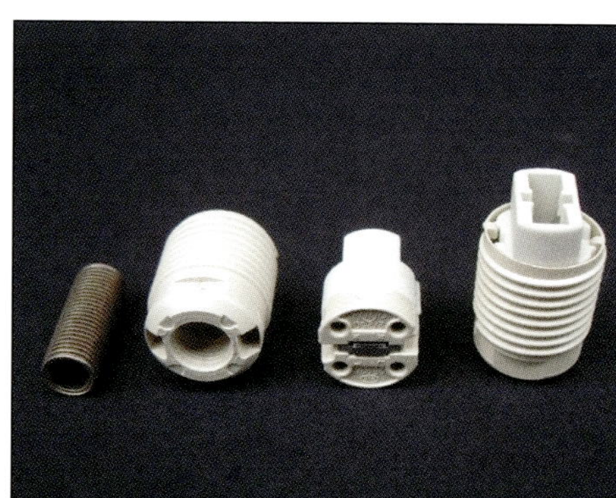

Lampholder for G9-based lamp.
Photos: Margaret O'Rorke.

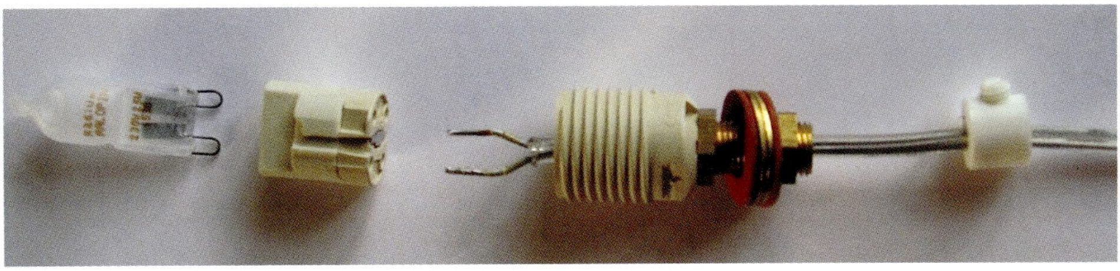

The lamp, washers, nuts & bolts and threaded tube needed for fixing.

each other. The flex must be secured with a flex restrainer before it leaves the lamp fitting. The position of the holes for fixing the lamp holder must be thought through before making the ceramic, and you should remember to allow for access to get the lamp in and out.

A small high-speed drill with diamond drill bits, which can be bought in model-making shops, is very useful for adjusting the size of fired holes, etc.; a flexible shaft attached to the drill is needed, which makes it safer to hold the ceramic under running water so as to prolong the life of the drill bits.

The G9 lamps have 2000 hours' lamp life. However, they do produce a lot of heat, so any ventilation holes you can add underneath will help to lengthen the life of the lamp.

The 240-volt (110-volt in the USA and Canada) 60-watt linear halogen K34 with R7s base is a useful lamp and holder if you need a bright low-fitting light source. It is also dimmable. Make sure you buy the correct dimmer switch for the number of watts installed. Remember that the heat output is high, so try to create as much ventilation through the light as possible. Upward convection flows will remove large amounts of heat, whereas a closed top will simply contain all the warmth.

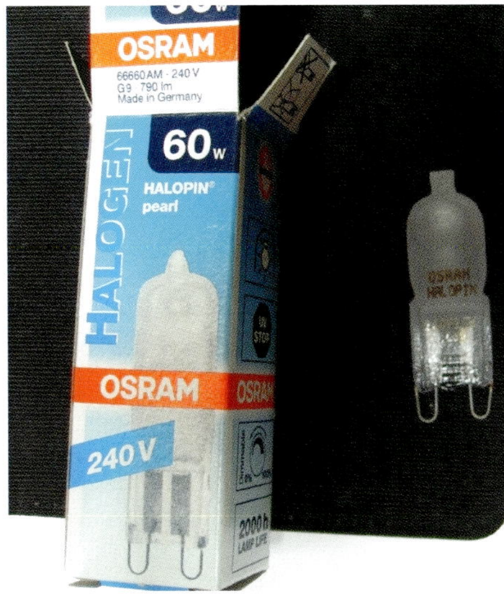

The 240-volt capsule lamp with G9 base.

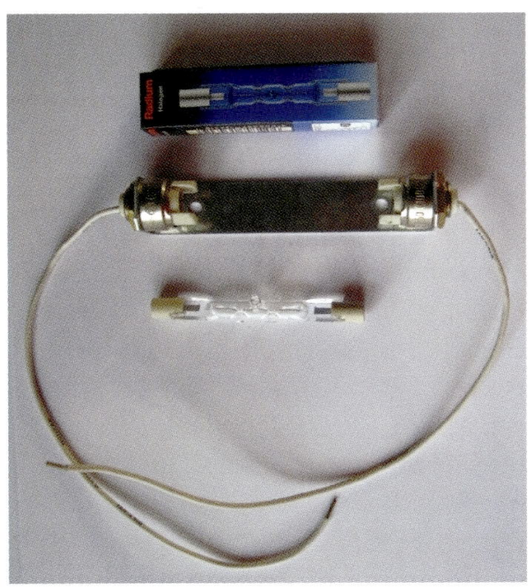

R7s lamp-holder assembly and K34 lamp.
Photos: Margaret O'Rorke.

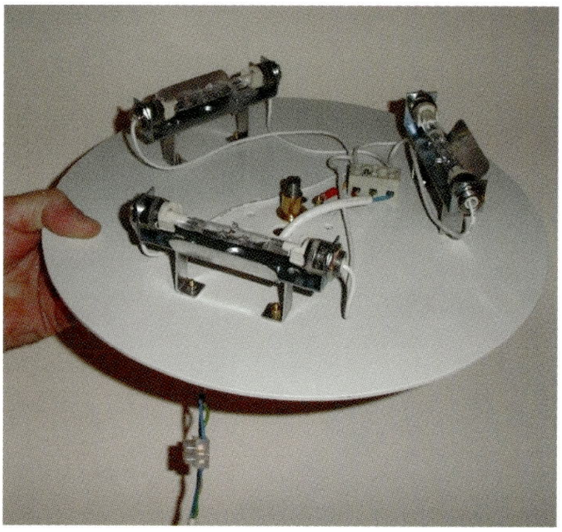

Example of the wired and linked lamp holders for this ceiling light. 12 cm (4¾ in.) long, 3.5 cm (1⅜ in.) high, with glass shield.

Cast ceiling light. Note that all the engineering is unseen.

The fluorescent tube with separate electronic ballast, whether compact or linear, is a neat slimline system for larger installations. Remember always to use triphosphor tubes, as the colour rendition is so much better. Follow the ballast wiring instructions sheet carefully. Remember to use flex restrainers on the mains power lead, and make sure you order the ballast that is compatible with your chosen lamp. It is easiest to use solid-core cable to wire the ballasts and lamp holders. If in any doubt ask a qualified electrician.

The slimline lamp is available in a warm or cool tone and does not generate much heat. The lamps are available from 14 to 35 watts (T5/16 mm tube) or from 15 to 70 watts (T8/25mm tube).

Fluorescent lamps, whether compact or linear, are energy-saving, i.e. very efficient in producing a large amount of light for a comparatively small amount of power. They also have a long life, making the need for replacement of lamps less frequent.

I have given an example of how the wiring of the lamps to the ballast can be done. Adding coloured tape helps you to identify the correct wiring (see the following sequence, p.95).

I have not yet been able to find a bright enough source of light-emitting diodes (LEDs) in the colour range that I want to use, though the technology for LEDs is developing rapidly now and I intend to continue researching the possibilities. This may become a major light source for the ceramic artist.

Fibre optics are a costly investment to experiment with. I have not yet found the diffused spread of light required for the type of work I am interested in developing, though of course things could change. The information offered in this chapter is intended to introduce and highlight some of the technical aspects involved in introducing electrical lighting elements to ceramic forms. It will, I hope, help the ceramic artist to research and experiment with their ideas. For expert advice, and before selling your work, please read the following safety section of this chapter, and have your work checked by a qualified electrician.

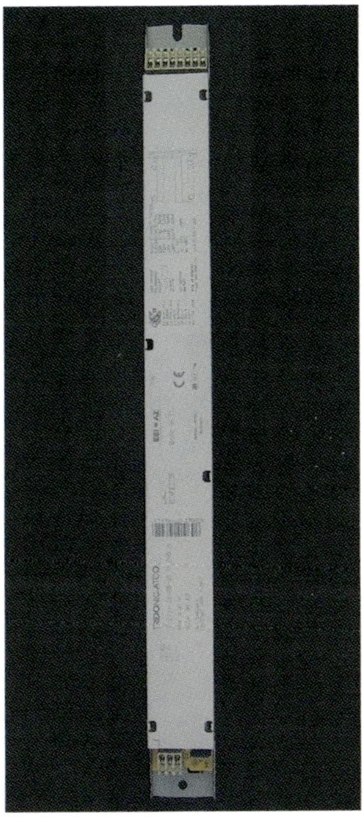

Ballast.

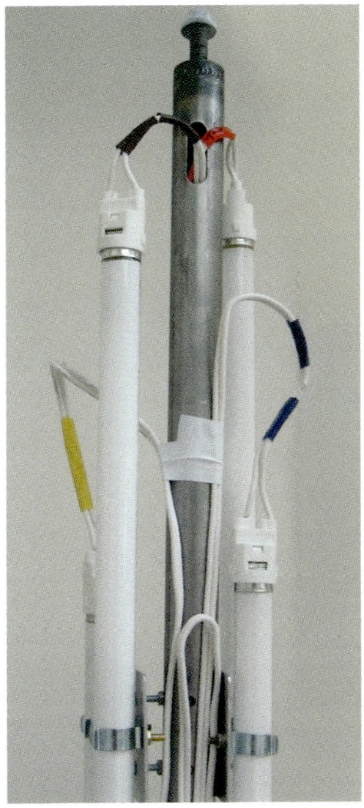

Top of pole.

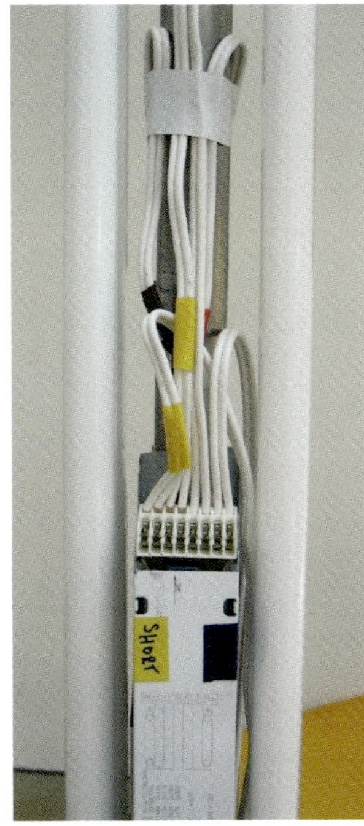

Top of wiring ballast.

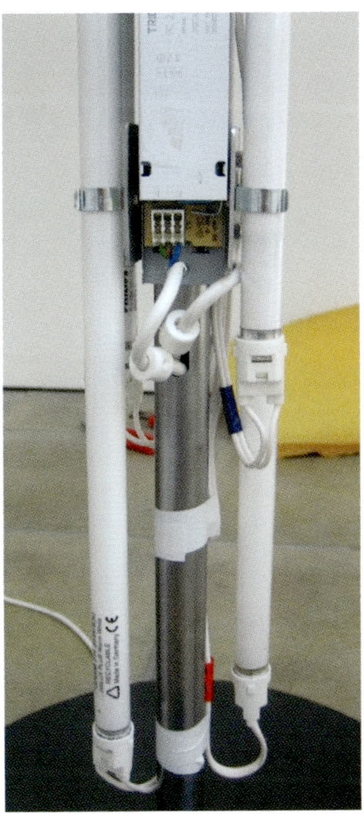

Bottom wiring of ballast.

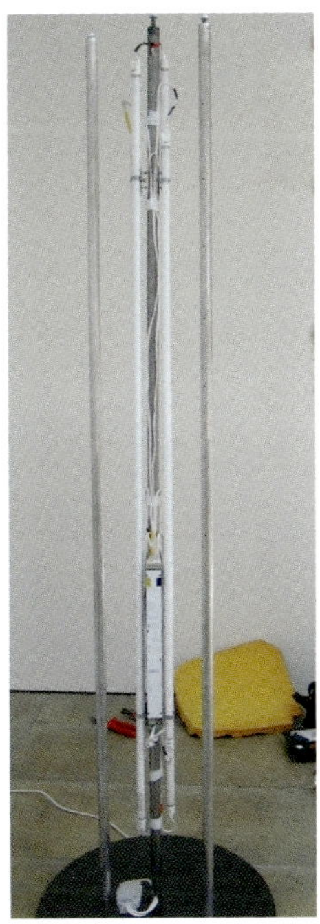

Lamps and wiring.

Porcelain wings attached.
Photos: Margaret O'Rorke.

Safety information

The following information was sent to me by Patrick Saunders, a tutor on the electrical course at Connaught Compliance (formerly known as NECTA), an electrical safety and compliance organisation. In response to my request for safety information for different parts of the world, he wrote:

'Obviously, these countries have similar but different electrical safety standards and governing bodies regarding the approval of electrical appliances. However, different bodies and standards may apply to various market sectors, i.e. domestic, commercial and industrial. The designer and any subcontracted manufacturer are obliged to ensure that the specific electrical and other relevant safety standards would be met for your appliance(s) to be approved in that particular country and intended market sector.'

'For example, in the **UK** the main bodies are the HSE (Health & Safety Executive, although this closed in 2009, their publications are available online at www.hse.gov.uk), BSI (British Standards Industry), IEE (Institution of Engineering and Technology), ESC & ASTA-BEAB (a global product and systems certification service to manufacturers of electrical products). Please note that some of these are 'state bodies' and some are 'notified bodies'. In **Europe** we have CENELEC (European Committee for Electrotechnical Standardization, which prepares voluntary electrotechnical standards that help develop the Single European Market/ European Economic Area for electrical and electronic goods and services, removing barriers to trade and cutting compliance costs), and the IEC (the international standards and conformity assessment body for electrotechnology), and also some European directives such as the LVD (Low Voltage Directive).

The **US** does not have a national safety body or institute as such, but they have ANSI (the American National Standards Institute), OSHA (Environment, Health and Safety Online, giving programs, assistance, risk assessment and regulations to download) and the UL (creating standards for safety). Please note that all the previous references are non-exhaustive.

'Having mentioned all of the above, with respect to the type of electrically powered sculptures that you create, it should not be too difficult to comply with the basic standards and requirements due to their relative simplicity. Many of the key elements with respect to electrical appliance safety are common; therefore, I would suggest that it may be prudent to look at some electrical training for perhaps yourself or anyone that assists you in your work. This may help you in ensuring that the appropriate choice of "materials", implementation and electrical-safety testing would be adopted. Below I have listed two short electrical training courses that should be of benefit.'

Course title	Duration
1. Basic Electrical Skills	3 days
2. IEE Code of Practice for the Service & Repair of Electrical Equipment (C&G 2377)	2 days

These courses are run by Connaught Compliance (see electrical support services on p.97 opposite for details).

Some basic electrical safety: How to wire a 3-pin plug

A plug is correctly wired as follows: the green/yellow insulated core goes to the earth terminal (top); and the brown insulated core goes to the live terminal. The live terminal can be identified by the fuse. The fuse and the switch should always be connected to live as it is the live which supplies electrical energy to the device. If the fuse and switch were connected to neutral, the appliance could appear to be off, yet could still be connected to live. The blue insulated core goes to the neutral terminal.

You can also access this information on the internet at: www.allsands.com/potluck4/plugthreepinw_ukw_gn.htm

For those working in the UK and USA, *Basic Wiring* by Ken Sidey (editor) (Stanley Books, published by Meredith Books, 2002) could be useful. It is available in both the UK and USA.

Electrical information available on the internet

In the UK, a good website for basic electrical safety information is this one, helping people to problem-solve: www.ccohs.ca/oshanswers/safety_haz/electrical.html#_1_8

The useful American version is a simple website with DVD explaining basic home wiring: www.electrical-online.com/DVDBasicHomeWiring.htm

Electrical support services

UK

Connaught Compliance Electrical Services
+44 (0)1709 360621
www.nectasafety.co.uk
For electrical training programmes, electrical inspection and testing, and engineering support services.

The Lighting Industry Federation
www.lif.co.uk
Professional advice, training and services.

USA

American Lighting Association
www.americanlightingassociation
Educational programs and services.

www.electrical-online.com/DVDBasicHomeWiring.htm
A reference for basic home wiring.

For general electrical suppliers, please see p.110.

Chapter 6

Fountains

Combining running water with translucent lit porcelain is an intriguing and exciting concept with fantastic potential. Imagine the fine porcelain form lit up, with the water reflecting sparkling light. First, a few questions to ask yourself: Where is the setting for the fountain? Is it to be an interior or exterior fountain? How will the water flow over the forms? What will the scale be? The questions posed when attempting to create a lit porcelain water feature demand that the ceramic artist expands their sensory awareness yet further.

The basic discovery that water poured into a spoon splashes out and that, when the spoon is turned over, it flows over and down without splashing, is a useful fundamental to remember; it took me time before I figured it out. We have discussed how the light must be within or behind the porcelain in order to reveal the translucency, and that when directed onto it the translucency is masked. So, in this instance, you should ask yourself what course will the water take when flowing over the work and what will contain the water. Porcelain is an excellent material for outdoor use; but if, because of its translucency, fine porcelain is part of the concept, this will be more delicate and would perhaps be better suited for interiors.

To begin with, for the interior designer, architect and ceramic artist, the space the feature is intended to inhabit must be the first consideration. Will the lights be immersed in the water or sealed from it? Will they be above, below or behind the ceramic? How can the pump remain unseen while still circulating the water efficiently? What sounds might the water make?

On the internet are illustrations of fountains using ceramics with light, but I have not been able to find any examples using the translucent qualities of porcelain or bone china combined with water and light.

If as an artist you are embarking on your first lit water feature, you will find it helpful to pay a visit to your local aquarium or garden centre to familiarize yourself with some of the technicalities involved. Discuss with them what you want to do. There is also a great deal of information on the internet (see the relevant websites in the list of suppliers at the end of this book). You might decide to buy a ready-made system and adapt it, though personally I have never wanted to work with the constraints this would impose.

A good pump plus the lights are expensive, together with the appropriate waterproof electrical junction boxes and the housing for it all. So my suggestion is to begin with the simplest possible set-up so that you can experiment with bringing the three aspects of the work together on a small scale. To begin with, making a model using water-resistant materials and then running water over to help you to see how it might work. It

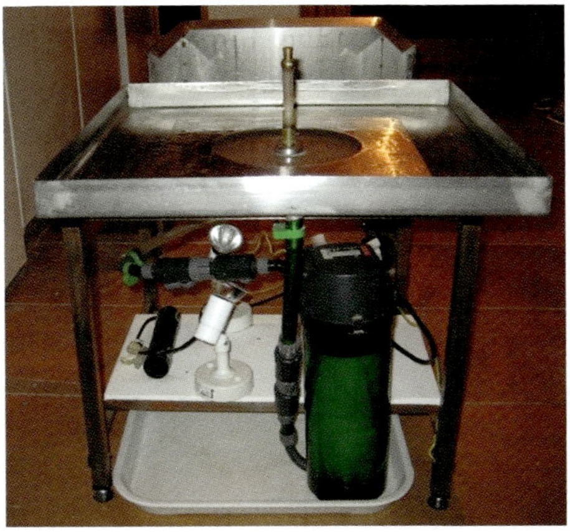

1 Light isolated from water. *Photo: Margaret O'Rorke.*

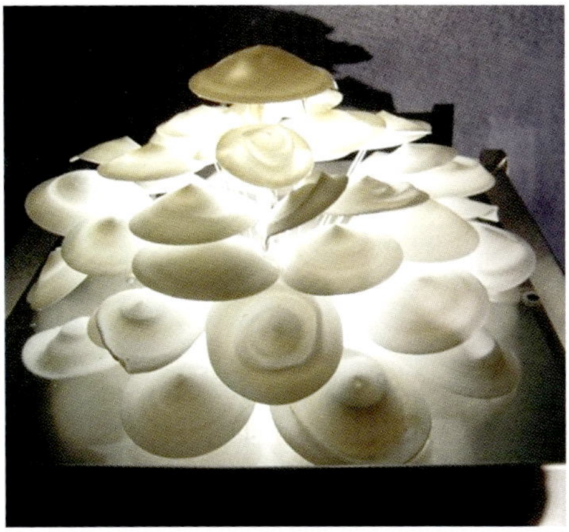

2 One of the four corners of the water features. *Photo: James Crabbe.*

will also highlight some of the details to be resolved before you begin making and firing ceramic pieces.

To help you get started I have illustrated an example of one of my own fountain projects, in order to show the process. This has been a learning curve, which could lead on to much larger installations in the future.

Photo no.1 illustrates a stainless-steel stand housing a 12-volt transformer connected to three adjustable spotlights. These shine from beneath through a circular, sanded Perspex sheet. It is sealed and bolted to the stainless-steel tray. An Eheim pump is connected to sealed inlets, pumping the water up to a spray head at the end of a flexible tube and back to the pump. The supporting frame, seen at the back of the stand, is surrounded by folded stainless-steel sides that hang from the rim of the tray. The thrown porcelain hats are suspended above the light on stainless-steel rods (see photo no.2, top right). These four small fountains formed the corner pieces of the installation shown in photo no.4 (overleaf). The tall centre piece was a pyramid of thrown,

3 The centrepiece of the fountain, 1995. Ht: 57 cm (22½ in.). *Photo: Arnold Borgerth.*

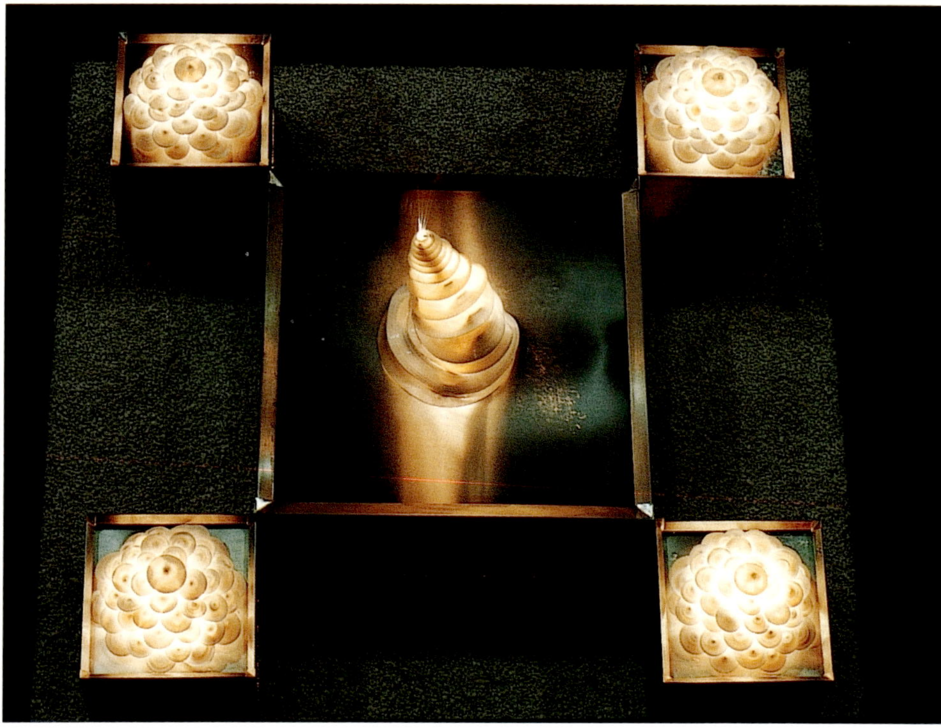

4 ABOVE Water feature installation centrepiece 3 m x 3 m (9 ft 8 in.) at Olympia Fine Art and Antiques Fair, 1995. *Photo: John Woroniecki.*

1 Jackson Li in 2007, observing the translucency of the first cast fine-porcelain bamboo forms for the fountain at Sanbao, Jingdezhen, 2007. *Photo: Ric Swenson.*

cut and joined porcelain rings.

Together they made the centrepiece at the entrance of the Fine Art and Antiques Fair at Olympia in London in 1995.

In spring 2008, Lea Bly came to Oxford as my assistant and helped to create the machete prototype for a new bamboo water feature, the idea for which had begun in China the year before in collaboration with Jackson Li at Sanbao, Jingdezhen (see photo no.1, left).

In Oxford we first cut up garden bamboo sticks and experimented with different ways of stacking them, and watched the way water flowed over them. Lea made moulds and we slipcast the bamboo rods (photo no.3, opposite). This enabled me to see the way the water might flow and what the light would do. It also meant I could send photos to China to illustrate my idea ahead of time (photo no.4, p.102).

When I arrived in Sanbao again in August 2008, preparations had been made and we were able to get straight down to work. Liqiang, my assistant, went up into the forest to cut the widest bamboo trees he could find, from which he made the moulds (see below). Shar helped (photo no.6), and together they cast and fired 240 pieces.

When the pieces had all been fired (photo no.7), a stand was made from bamboo to support the work and to house the underwater lights and pump (photo no. 9 and 10). The frame was made and installed in an indoor conservatory pond. Mr May got the whole thing working (photo no.12), and we began to install the cast porcelain bamboo pieces (photo no.13). But our first attempt at stacking the pieces did not work well, because they were of different proportions to the model (see photo no.4, p.102).

There were naturally a lot of hitches. One of the submersible lights I had

2 ABOVE Lea Bly testing the idea.

3 BELOW Plaster moulds for slipcasting. *Photos: Ric Swenson.*

4 Machete porcelain.

5 Liqiang making the bamboo mould.

6 Shar helping Liqiang slipcasting.

brought from the UK flooded. It was Mr Huang, the owner of a large Chinese lighting company and a friend of Jackson's, who quickly replaced it for us. This demonstrated how important it is for the artist to find an electrical technologist who is willing to support a creative project (if you are working to a specific commission or you plan to sell the feature, it is essential to seek professional advice). When that was fixed we re-designed the fountain, and it was carefully put together by Liqiang and a helper (photo no.14), while I drilled holes in some of the pieces to form a lit wall piece to accompany the fountain (photo no.17).

I have included a number of web sites in the list of suppliers at the end of the book with information that could be of interest to potential makers. The Fountain Society (see suppliers' list p.111) has a website with help and advice, and also provides a useful booklet with information about fountains.

7 RIGHT TOP Kiln load fired to 1330°C (2426°F).

8 RIGHT Shar rubbing down the fired porcelain.

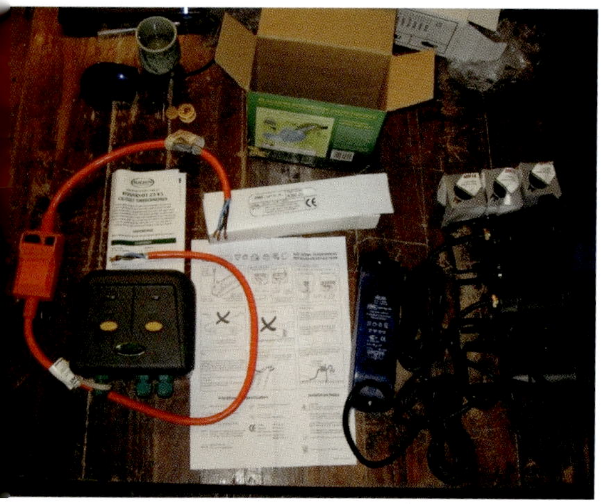

9 Waterproof electrical fittings (from left to right): circuit breaker, insulated junction box, 12-volt transformer, and three submersible halogen spotlights.

10 Submersible Eheim pump, water-pressure control taps.

11 Installing the bamboo frame in the pool.

12 Mr May got it all working.

13 I began to lay the pieces but they did not look right.

14 Liqiang and his assistant Shar lay them another way, in a more formal arrangement.

15 LEFT The finished piece.

16 RIGHT Detail of the finished piece.

17 BELOW Translucent porcelain bamboo wall installation. 100 x 170 cm (39 x 67 in.).

Conclusion

In Una Lee's words 'Light is everywhere'. Moholy-Nagy the Bauhaus designer who encouraged the development of lighting design said, 'Allow function to determine design'.

My hope is that this small handbook has introduced some of the possibilities for a creative craftsperson wanting to develop lighting using clay, whether in the studio or as a commercial product for industry. I hope it has challenged them to make works that reflect the nature of the material, their creative imagination and their understanding of how the quality of light can enhance people's lives.

Throughout history it has been the role of the craftsperson to communicate their character through their work, giving sustainable joy to future generations through their creative craft. The current climate of fast-moving communication, the balance of values between art and craft, together with the mass production of disposable goods, has somewhat clouded the true value of the art of craftsmanship. So it is my feeling that we who are hardworking and privileged to be at the cutting edge of the art/craft arena, need to establish a clear voice, to help those who are part of our current political, educational, social and commercial world to understand the need humanity will always have for the devoted, tangible creativity of the craftsperson.

I hope architects, interior designers and lighting companies will embrace the quality of light that translucent porcelain or bone china can impart to a space, which change the atmosphere of that space. I hope that they will engage a creative ceramic craftsperson with knowledge, skill and experience of the material at the earliest planning stage so as to give time for a lit ceramic project to evolve.

As the technology for electrical lighting continues to develop rapidly, it can be the role of the ceramic artist to respond by ensuring that the balance between the technology and the quality of light needed to create a relaxed, calm and peaceful atmosphere is maintained both in the home and in the workplace. Experimenting with fibre optics and LEDs as technology develops will give new possibilities for combining translucent porcelain within the realm of interior lighting design. Creating translucent lit water features may also be an exciting future.

Porcelain is the strongest, most durable material when high-fired; it is used on pylons in the form of electrical insulators. The finer the clay, the more translucent it becomes, which also makes it more fragile under impact. All these aspects must be addressed. Combining it with water can create a form of protection, a physical barrier between the observer and the lit feature. For large-scale installations it is necessary to incorporate multiple ceramic parts. Remember that the simplest solution is often the best but frequently the hardest to find.

The lighting and ceramic industries have not yet recognised the full potential that porcelain or bone china can hold for lighting products, both in the home and for large-scale interiors. Ceramic is an ideal material to use for lighting. It is virtually indestructible except under impact, does not transmit electricity, and is antistatic, making it less inclined to attract dust. The artist can bring new methods of developing commercially produced original ideas, helping to retain, sustain and revive a very skilled industry that is rapidly diminishing and being forced into liquidation.

Contemporary ceramic artists are free to explore, research and develop far-reaching concepts, and have the opportunity to find subtle ways of using the natural material in conjunction with light that can enhance people's lives. It was, after all, this freedom that led early man to first develop ways of using clay and fire to light up darkness when the sun went down.

Acknowledgements

The invitation from my editor Alison Stace to write this book was a surprise. It has forced me to use words instead of clay to convey my thoughts and ideas to a new generation of potters. I thank her for her patience and help in its production.

I am grateful to the Arts Council for two research and development grants which have enabled me to explore at first hand some industrial processes necessary in order to develop designs that can be manufactured.

On the home front, the encouragement from my partner, Professor James Crabbe, who persuaded me to take up Alison's offer, and without whose inspiration I would never have written the book, together with the generosity with which Jan Scriven has guided me to make sense of the words that I used, have been priceless. The support and encouragement from my family has given me the freedom to explore some of the exciting possibilities of porcelain, light and water, and I feel very privileged.

It has been an interesting adventure to leave my studio and seek out around the world some of the creative makers who are using translucent porcelain or bone china with light. I thank each one of the artists represented here for giving me an insight into their making processes.

I was trained in one era of ceramic art, from 1954–59, and have been productive since 1980 in a very different one. My teachers, John Williamson, Dick Kendall, Lucie Rie, Nigel Wood and others, together with my ceramic friends Joanna Constantinidis, Sebastian Blackie, Ryoji Koie, Ann Linnemann, Neil Brownsword and many others, have been seminal in my creative life. I am also very grateful to Anita Besson for introducing my work to a wider audience.

List of suppliers

Clays UK

Valentine Clays Limited (UK)
The Sliphouse
18–20 Chell Street, Hanley
Stoke-on-Trent, ST1 6BA
Tel: +44 (0)1782 271200
Fax: +44 (0)1782 280008
www.valentineclays.co.uk/
Suppliers of Audrey Blackman porcelain.

Potclays Ltd
Brickkiln Lane
Etruria, Stoke-on-Trent
Staffordshire, ST4 7BP
Tel: +44 (0)1782 219816
www.potclays.co.uk
Suppliers of various porcelain bodies including Southern Ice, and also stock bone china bodies.

Clays USA

Ceramic Store Inc.
1002 West 11th Street
Houston TX 77008
Tel: +1 (713) 864-6442
store.ceramicstoreinc.com
Porcelain clays online.

Laguna (USA)
Laguna-Ohio
61020 Leyshon Drive
Byesville, OH 43723
Toll-free: +1 (800) 762-4354
Local: +1 (740) 439-4355
www.lagunaclay.com

Laguna-California
14400 Lomitas Avenue
City of Industry, CA 91746
Toll-free: +1 (800) 4-LAGUNA
Local: +1 (626) 330-0631
www.lagunaclay.com

Electrical suppliers & resources (UK)

www.preferredlightinginc.com
www.acornlighting.com
Specialists in garden lighting.
www.andythornton.com/products.asp/subcatID/29/led-lights.htm
Lights (worldwide network).
www.energysaverlamps.co.uk
A good online source of lamps.
www.thegardenlightingshop.co.uk
Outdoor waterproof lamps.
www.litefuzion.com
Charger for lots of nitelite LED candles.
www.lampsandlights.co.uk
Variety lighting hardware
www.thelightbulb.co.uk
Specialists in bulbs.
uk.rs-online.com/web
A quick and comprehensive supplier for basic light fittings, tools and flex, etc. providing you know what you want.
www.specialistlampfittingsupplies.co.uk/components.html
Specialist lamp-fitting suppliers.

USA & Worldwide
www.energystar.gov/lighting
Energy-efficient lighting info & supplies.
www.andythorntonusa.com
Lights (worldwide network).
www.powerfilmsolar.com
Solar-powered chargers.
www.purdyelectronics.com
LEDs, fibre optics, etc.

www.chinagiftware.com/ledcandle/
LED candles direct from China.

Other useful organisations (UK)

The Fountain Society
Hon. Secretary, Ian Hay-Campbell
High House, Bucknell
Shropshire, SY7 0AA
www.fountainsoc.org.uk

AIR
a-n The Artists Information Company
First Floor, 7–15 Pink Lane
Newcastle upon Tyne, NE1 5DW
www.a-n.co.uk
General enquiries: info@a-n.co.uk

Bibliography and Resources

Caspall, John, *Making Fire & Light in the Home pre-1820* (Woodbridge, Suffolk, Eng.: Antique Collectors' Club, 1987). (out of print)

Cox, James A., *A Century of Light* (New York: The Benjamin Co., 1979) (basically electric lighting and Edison – out of print)

Early Lighting: A Pictorial Guide (The Rushlight Club, second edition, 1979).

Faraday, Michael, *The Chemical History of a Candle* (J.M. Dent & Son, 1920).

Paton, J. *Lamps: A Collector's Guide* (Souvenir Press Collectors' Series, 1979).

Robins, F.W., *The Story of the Lamp* (Oxford University Press, 1939, repr. 1970).

Sidey, Ken (editor), *Basic Wiring* (Stanley Books, 2002).

Szentléky, T. *Ancient Lamps* (Argonaut, tr. 1969).

Thwing, Leroy L., *Flickering Flames* (Rutland, VT: Charles E. Tuttle Co., 1958), (available on Amazon.com).

www.worldwideflood.com/.../oil_lamps.htm

Good history of oil lamps and how to make them.

www.prm.ox.ac.uk

Pitt Rivers Museum, Making Light Work, an exhibition of lighting including the work of Margaret O'Rorke, January 1987.

Websites/contacts of contributors

Max von Barnholt	www.vbklighting.co.uk
Paula Bastiaansen	www.paulabastiaansen.com
Beverly Bigham	Tel: 0207 731 0805
Jeremy Cole	www.jeremycole.net
Ilena Finocchi	www.ilenaf.com
Mia Göransson	www.miaegoransson.com
Ian Johnston	www.theartark.com/johnstonfischer.html
Kurokawa Kenkichi	www.kenkichi.org
Una Lee	email: iuna1222@yahoo.co.kr
Inkeri Leivo	www.artarabia.fi
Lilach & Ron Lotan	www.ceramic-art.ca
Tomasz Niedziólka	email: naczynie@wp.pl
Margaret O'Rorke	www.castlight.co.uk
Liisa Rissanen	email: lrissane@luukku.com
Mark & Emma Welsh	www.itemhouse.co.uk

Index

(**Bold** indicates images)